Algebra 1
Concepts and Skills

Larson Boswell Kanold Stiff

California Standards Key Concepts Book

This book contains a review of pre-course skills, key standards support including teaching and practice, and special topics.

McDougal Littell
A HOUGHTON MIFFLIN COMPANY

Evanston, Illinois • Boston • Dallas

Acknowledgement

We acknowledge with special thanks the contributions to the planning, writing, and reviewing of the Special Topics section on pages T1–T25 by:

Special Contributor

Richard Askey
Professor of Mathematics
University of Wisconsin
Madison, Wisconsin

ISBN: 0-618-07876-2

6789-DWI-06 05 04 03 02

Contents

Contents *continued*

Part 3 Special Topics

Part 1 Pre-Course Review

Diagnostic Tests

Topic 1: *Working With Fractions*

Lesson 1

Write each fraction in simplest form. If it is already in simplest form, write simplest form.

1. $\dfrac{20}{64}$

2. $\dfrac{27}{45}$

Lesson 2

Write each mixed number as an improper fraction.

3. $5\dfrac{3}{5}$

4. $1\dfrac{24}{25}$

Write each improper fraction as a mixed number.

5. $\dfrac{20}{11}$

6. $\dfrac{82}{7}$

Lesson 3

Find each sum or difference. Write each answer as a fraction or mixed number in simplest form.

7. $\dfrac{15}{16} + \dfrac{1}{4}$

8. $1\dfrac{3}{7} + 2\dfrac{1}{2}$

9. $\dfrac{2}{3} - \dfrac{5}{12}$

Lesson 4

Multiply or divide. Write each answer as a fraction or mixed number in simplest form.

10. $\dfrac{2}{5} \times \dfrac{7}{10}$

11. $\dfrac{6}{7} \div \dfrac{5}{14}$

12. $1\dfrac{1}{4} \div \dfrac{3}{4}$

Topic 2: *Rates, Ratios, and Percents*

Lesson 1

Write each decimal as a percent and as a fraction or mixed number in simplest form.

13. 0.81

14. 1.7

Write each fraction or mixed number as a percent and as a decimal.

15. $2\dfrac{9}{20}$

16. $\dfrac{7}{16}$

Write each percent as a decimal and as a fraction or a mixed number in simplest form.

17. 4%

18. 105%

Lesson 2

Write each ratio in lowest terms.

19. 5:10

20. 18:15

Find the unit rate.

21. 165 miles in 3 hours

22. $18,000 in 12 months

Lesson 3

Find the percent of the number.

23. 40% of 25

24. 12% of 32

Find the percent.

25. 27 out of 60

26. 3 out of 30

Topic 3: *Integers*

Lesson 1

Find each sum.

27. $-12 + 15$

28. $-25 + (-41)$

29. $-24 + 21$

Lesson 2

Find each difference.

30. $-15 - 30$

31. $-28 - (-20)$

32. $24 - 40$

Lesson 3

Find each product or quotient.

33. $8(-5)$

34. $-48 \div (-6)$

35. $(-3)(7)(-2)$

Lesson 4

Evaluate each expression.

36. $15 - 3 \cdot 4$

37. $\dfrac{70 \div (8 - 1)}{1 + 2^2}$

Lesson 5

Evaluate each expression.

38. $20t - 12$ when $t = 3$

39. $(x + 5)^2$ when $x = 7$

Topic 4: *Measurement*

Lesson 1

Complete.

40. 1.25 mi = __?__ yd

41. 6 in. = __?__ ft

42. 8 lb = __?__ oz

Lesson 2

Complete.

43. 400 mL = __?__ L

44. 100 kg = __?__ g

45. 2 m = __?__ cm

Lesson 3

46. Find the perimeter of a rectangle with a length of 25 cm and a width of 20 cm.

47. Find the circumference of a circle with diameter 48 mm. Use 3.14 for π.

Lesson 4

48. A square with sides of length 10 m is cut in half along one of the diagonals. Find the area of one of the triangles formed.

49. Find the area of a rectangle with a length of 2 ft and a width of 0.5 ft.

50. Use the formula $A = \dfrac{1}{2}(b_1 + b_2) \cdot h$ to

find the area of the trapezoid shown.

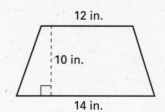

12 in.

10 in.

14 in.

Topic 5: *Working with Data*

Lesson 1

For Exercises 51 and 52, use the bar graph.

51. Estimate the average cost of operating a sport utility vehicle.

52. Which types of vehicles cost 45 cents/mile or less?

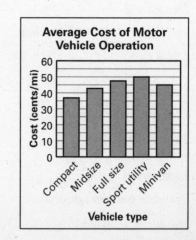

Average Cost of Motor Vehicle Operation

Cost (cents/mi)

Vehicle type

Compact, Midsize, Full size, Sport utility, Minivan

53. Draw a line graph to display the data in the table.

Average High Temperature (°F) in Portland, Oregon											
Jan.	Feb.	Mar.	Apr.	May	June	July	Aug.	Sept.	Oct.	Nov.	Dec.
45	51	56	61	67	74	80	80	75	64	53	46

Lesson 2

For Exercises 54 and 55, use the circle graph.

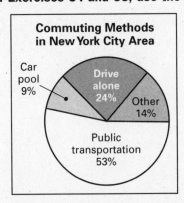

54. If the population of the New York City area is about 20 million, how many people take public transportation?

55. If the population of the New York City area is about 20 million, how many more people drive alone than carpool?

Lesson 3

For Exercises 56 and 57, use the two graphs below that illustrate the same data.

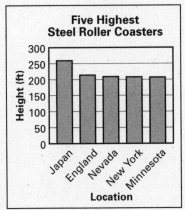

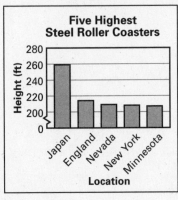

56. Which graph suggests that the Roller Coaster in Japan is more than twice as high as the Roller Coaster in England? Is this impression correct?

57. Why do these graphs give such a different visual impression?

Topic 1 Warm-ups

Standardized Testing Warm-Ups

1. Which one of the following is *incorrect*?

   A $150 \div 15 = 10$

   B $150 \div 25 = 6$

   C $30 \times 5 = 150$

   D $75 \times 2 = 140$

2. Which one of the following is *incorrect*?

   A $4 \times 9 = 36$

   B $2^2 \times 3^2 = 36$

   C $6(2) = 36$

   D $2 \times 2 \times 3 \times 3 = 36$

Homework Review Warm-Ups

Divide.

3. $32 \div 16$

4. $42 \div 14$

5. $125 \div 5$

6. $320 \div 8$

7. $420 \div 60$

8. $125 \div 25$

Topic 1 — Adding and Subtracting Fractions

GOAL

Add and subtract fractions that have like and unlike denominators.

Ruby is helping at a bake sale. There are parts of two pies left. How can she tell if there is more than a whole pie left to sell?

UNDERSTANDING THE MAIN IDEAS

You can add and subtract fractions. To add two fractions with the same, or like, denominators, add the numerators. Write your answer in simplest form.

Example 1

Find each sum.

a. $\dfrac{1}{9} + \dfrac{5}{9}$

b. $\dfrac{6}{10} + \dfrac{9}{10}$

 Solution

a.

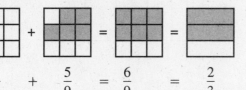

$$\frac{1}{9} \;+\; \frac{5}{9} \;=\; \frac{6}{9} \;=\; \frac{2}{3}$$

b.

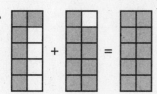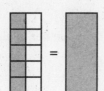

$$\frac{6}{10} \;+\; \frac{9}{10} \;=\; \frac{15}{10} \;=\; 1\frac{1}{2}$$

To add fractions with different, or unlike, denominators, you need to write equivalent fractions that have a common denominator, which means the fractions are in the same size pieces. After you add, write your answer in simplest form.

Example 2

Find each sum.

a. $\dfrac{7}{10} + \dfrac{1}{2}$

b. $\dfrac{2}{3} + \dfrac{1}{4}$

■ Solution ■

a. Since $\dfrac{1}{2} = \dfrac{1 \cdot 5}{2 \cdot 5} = \dfrac{5}{10}$, use 10 as the common denominator.

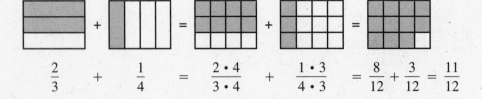

$$\dfrac{7}{10} \;+\; \dfrac{1}{2} \;=\; \dfrac{7}{10} \;+\; \dfrac{5}{10} \;=\; \dfrac{12}{10} = 1\dfrac{2}{10} \;=\; 1\dfrac{1}{5}$$

b. Because 3 and 4 have no common factor, write equivalent fractions using the common denominator 3 • 4 or 12.

$$\dfrac{2}{3} \;+\; \dfrac{1}{4} \;=\; \dfrac{2 \cdot 4}{3 \cdot 4} \;+\; \dfrac{1 \cdot 3}{4 \cdot 3} \;=\; \dfrac{8}{12} + \dfrac{3}{12} = \dfrac{11}{12}$$

Find each sum. Write each answer as a fraction or mixed number in simplest form.

1. $\dfrac{9}{20} + \dfrac{7}{20}$

2. $\dfrac{5}{6} + \dfrac{5}{6}$

3. $\dfrac{1}{10} + \dfrac{7}{20}$

4. $\dfrac{3}{4} + \dfrac{7}{12}$

5. $\dfrac{1}{3} + \dfrac{1}{4}$

6. $\dfrac{1}{2} + \dfrac{2}{3}$

To subtract fractions, you also need to write equivalent fractions that have a common denominator. Remember to write your answer in simplest form. When dealing with mixed numbers, it is often helpful to convert to an improper fraction first.

Topic 1 *Warm-ups*

Standardized Testing Warm-Ups

1. Find the sum: $\dfrac{7}{15} + \dfrac{1}{5}$

    A $\dfrac{8}{20}$ **B** $\dfrac{8}{15}$ **C** $\dfrac{2}{3}$ **D** $\dfrac{4}{5}$

2. Find the difference: $1\dfrac{4}{9} - \dfrac{17}{18}$

    A $\dfrac{1}{2}$ **B** $\dfrac{5}{18}$ **C** $\dfrac{11}{2}$ **D** $\dfrac{11}{18}$

Homework Review Warm-Ups

Find each sum or difference. Write each answer as a fraction or mixed number in simplest form.

3. $2 - \dfrac{3}{8}$ **4.** $\dfrac{11}{2} + 2\dfrac{1}{3}$ **5.** $1\dfrac{5}{12} - \dfrac{3}{4}$

Jennifer ·b·

Topic 1

Multiplying and Dividing Fractions

GOAL

Multiply and divide fractions and mixed numbers.

You can use fraction operations when dealing with food recipes. For example, if you have only one measuring cup that holds $\frac{1}{3}$ cup and you need $3\frac{1}{3}$ cups of flour, how many times do you have to fill the measuring cup? You can use division to solve this problem.

Terms to Know	Example / Illustration
Reciprocals two numbers whose product is 1	$\frac{5}{6}$ and $\frac{6}{5}$ are reciprocals because $\frac{5}{6} \times \frac{6}{5} = 1$.

UNDERSTANDING THE MAIN IDEAS

To multiply fractions, multiply the numerators and multiply the denominators.

Example 1

Multiply $\frac{2}{3} \times \frac{6}{11} \times \frac{1}{2}$.

■ Solution ■

$$\frac{2}{3} \times \frac{6}{11} \times \frac{1}{2} = \frac{2 \times 6 \times 1}{3 \times 11 \times 2}$$ Multiply numerators. Multiply denominators.

$$= \frac{12}{66}$$ Simplify.

$$= \frac{12 \div 6}{66 \div 6}$$ Divide numerator and denominator by GCF, 6.

$$= \frac{2}{11}$$ Write fraction in simplest form.

Multiply.

1. $\frac{4}{5} \times \frac{15}{18}$

2. $\frac{21}{8} \times \frac{2}{7}$

3. $\frac{1}{3} \times \frac{3}{4} \times \frac{5}{9}$

To multiply any combination of fractions, mixed numbers, and whole numbers, first write all of the factors in fraction form.

Example 2

Multiply $3\frac{1}{2} \times 8$.

Solution

$$3\frac{1}{2} \times 8 = \frac{7}{2} \times \frac{8}{1} \qquad \text{Rewrite numbers as fractions.}$$

$$= \frac{7 \times 8}{2 \times 1} \qquad \text{Multiply numerators. Multiply denominators.}$$

$$= \frac{56}{2} \qquad \text{Simplify.}$$

$$= 28 \qquad \text{Write answer in simplest form.}$$

Multiply.

4. $3 \times \dfrac{5}{12}$ **5.** $\dfrac{3}{8} \times 4\dfrac{2}{7}$ **6.** $4 \times 2\dfrac{2}{5}$ **7.** $3\dfrac{2}{3} \times 1\dfrac{1}{2}$

To find the reciprocal of a number, write the number as a fraction. Then interchange the numerator and the denominator.

Example 3

Find the reciprocal of $3\frac{4}{9}$.

Solution

$$3\frac{4}{9} = \frac{31}{9} \qquad \text{Rewrite } 3\frac{4}{9} \text{ as a fraction.}$$

$$\frac{31}{9} \Rightarrow \frac{9}{31} \qquad \text{Interchange numerator and denominator.}$$

The reciprocal of $3\frac{4}{9}$ is $\frac{9}{31}$.

Find the reciprocal.

8. $\dfrac{7}{13}$ **9.** $3\dfrac{2}{5}$ **10.** $2\dfrac{12}{17}$ **11.** 4

To divide by a fraction or mixed number, multiply by its reciprocal.

Example 4

Divide $\dfrac{5}{8} \div 2\dfrac{3}{4}$.

■ **Solution** ■

$\dfrac{5}{8} \div 2\dfrac{3}{4} = \dfrac{5}{8} \div \dfrac{11}{4}$ Rewrite mixed number as fraction.

$= \dfrac{5}{8} \times \dfrac{4}{11}$ The reciprocal of $\dfrac{11}{4}$ is $\dfrac{4}{11}$.

$= \dfrac{20}{88}$ Multiply numerators. Multiply denominators.

$= \dfrac{5}{22}$ Write fraction in simplest form.

Divide.

12. $\dfrac{7}{5} \div \dfrac{5}{7}$ **13.** $18 \div \dfrac{9}{10}$ **14.** $5\dfrac{1}{4} \div \dfrac{7}{16}$ **15.** $1\dfrac{1}{4} \div 2\dfrac{2}{5}$

....
Spiral Review

Write each fraction as a mixed number or as a whole number.

16. $\dfrac{24}{7}$ **17.** $\dfrac{54}{9}$ **18.** $\dfrac{43}{6}$

Topic 2 ⬮ **LESSON 1** ⬮ *Fractions, Decimals, and Percents*

GOAL

Convert between fractions, decimals, and percents.

In 1990, about 51 of every 100 people in the United States were females. The number of females per 100 people in the United States can be written as a fraction $\left(\dfrac{51}{100}\right)$, a decimal (0.51), or a percent (51%).

Terms to Know	*Example / Illustration*
Percent a ratio comparing a number to 100 (Percent means "per hundred.")	$51\% = \dfrac{51}{100}$ The symbol % is read "percent."
Repeating decimal a decimal in which a single digit or a block of digits repeats without end	0.234234234. . . The three dots (called an *ellipsis*) indicate that the pattern of digits continues to repeat. (*Note:* The decimal number can also be written using a bar over the repeating digits: $0.\overline{234}$.)

UNDERSTANDING THE MAIN IDEAS

To write a terminating decimal as a fraction, use an appropriate power of 10 as the denominator and simplify. For example, if x is a decimal ending in tenths, rewrite x as $\dfrac{10x}{10}$ and simplify. To write a decimal as a percent, move the decimal point two places to the right and add a percent symbol.

Example 1

Write 0.25 as a percent and as a fraction in simplest form.

■ Solution ■

$0.25 = 25\%$ Move decimal point two places to the right.

$0.25 = \dfrac{25}{100} = \dfrac{1}{4}$ 0.25 is twenty-five hundredths. Simplify.

Write each decimal as a percent and as a fraction or mixed number in simplest form.

1. 0.5   **2.** 0.185   **3.** 1.32

To write a fraction as a decimal, divide the numerator by the denominator. If the remainder is not zero, and a digit or a block of digits repeats, the decimal is called a *repeating decimal*.

Example 2

Write each fraction or mixed number as a decimal.

a. $\dfrac{7}{8}$ **b.** $2\dfrac{5}{11}$

■ Solution ■

a.

$$
\begin{array}{r}
0.875 \\
8\,\overline{)7.000} \\
\underline{6\,4} \\
60 \\
\underline{56} \\
40 \\
\underline{40} \\
0
\end{array}
$$

So, $\dfrac{7}{8} = 0.875$

b.

$$
\begin{array}{r}
0.4545\ldots \\
11\,\overline{)5.000000} \\
\underline{4\,4} \\
60 \\
\underline{55} \\
50 \\
\underline{44} \\
60 \\
\underline{55} \\
50
\end{array}
$$

So, $2\dfrac{5}{11} = 2.\overline{45}$.

Write each fraction or mixed number as a decimal.

4. $\dfrac{5}{12}$ **5.** $1\dfrac{7}{9}$ **6.** $3\dfrac{4}{5}$

To write a fraction as a percent, rewrite it as a fraction with denominator 100, if possible. If the denominator is not a factor of 100, divide the numerator by the denominator. Then change the decimal to a percent by moving the decimal point two places to the right and attaching the percent symbol.

Example 3

Write each fraction as a percent.

a. $\dfrac{13}{25}$

b. $\dfrac{5}{8}$

■ Solution ■

a. Since 25 is a factor of 100, rewrite the fraction with a denominator of 100. Then give the percent.

$$\frac{13}{25} = \frac{13 \cdot 4}{25 \cdot 4} = \frac{52}{100} \;\rightarrow\; 52\%$$

b. Since 8 is not a factor of 100, first use division to change the fraction to a decimal and then change the decimal to a percent by moving the decimal point two places to the right.

$$\begin{array}{r} 0.625 \\ 8\overline{)5.000} \\ \underline{4\,8} \\ 20 \\ \underline{16} \\ 40 \\ \underline{40} \\ 0 \end{array}$$

$0.\,6\,2\,5 \rightarrow 62.5\%$

Write each fraction or mixed number as a percent.

7. $\dfrac{17}{20}$

8. $\dfrac{5}{8}$

9. $1\dfrac{4}{5}$

To write a percent as a decimal, move the decimal point two places to the left and remove the percent symbol.

To write a percent as a fraction, write the percent (without the percent sign) over 100 and simplify the fraction, if possible.

Example 4

Write each percent as a decimal and as a fraction or mixed number in simplest form.

a. 46%   **b.** 175%

c. 30.1%   **d.** 0.4%

■ Solution ■

a. $\%\Rightarrow 4\,6.\%\Rightarrow 0.46$

$46\% = \dfrac{46}{100} = \dfrac{23}{50}$

$30.1\% \Rightarrow 3\,0.1\% \Rightarrow 0.301$

Use decimal to write fraction.

$0.301 = \dfrac{301}{1000}$

b. $175\% \Rightarrow 1\,7\,5.\% \Rightarrow 1.75$

$175\% = \dfrac{175}{100} = \dfrac{7}{4} = 1\dfrac{3}{4}$

d. $0.4\% \Rightarrow 0\,0.4\% \Rightarrow 0.004$

Use decimal to write fraction.

$0.004 = \dfrac{4}{1000} = \dfrac{1}{250}$

Write each percent as a decimal and as a fraction or mixed number in simplest form.

10. 65%   **11.** 220%   **12.** 0.5%

. . . .
Spiral Review

Add or subtract. Write each answer as a fraction or mixed number in simplest form.

13. $\dfrac{1}{2} + \dfrac{5}{6}$ **14.** $1\dfrac{2}{7} + \dfrac{1}{3}$ **15.** $2\dfrac{7}{9} - 1\dfrac{1}{3}$

Topic 2 **Practice**

Write each decimal as a percent and as a fraction or mixed number in simplest form.

1. 0.45   **2.** 0.675   **3.** 1.2   **4.** 0.128

5. 0.06   **6.** 0.3   **7.** 0.76   **8.** 5.6

9. 0.44   **10.** 4.75   **11.** 0.001   **12.** 1.01

Write each fraction or mixed number as a decimal.

13. $\dfrac{6}{8}$ **14.** $\dfrac{7}{10}$ **15.** $\dfrac{1}{3}$ **16.** $\dfrac{13}{30}$

17. $8\dfrac{1}{4}$ **18.** $\dfrac{9}{16}$ **19.** $5\dfrac{37}{50}$ **20.** $\dfrac{3}{11}$

Write each fraction or mixed number as a percent. If necessary, round to the nearest tenth of a percent.

21. $\dfrac{4}{20}$ **22.** $\dfrac{1}{8}$ **23.** $\dfrac{6}{11}$ **24.** $\dfrac{5}{2}$

25. $\dfrac{9}{10}$ **26.** $5\dfrac{3}{4}$ **27.** $\dfrac{7}{15}$ **28.** $\dfrac{1}{3}$

29. $\dfrac{15}{40}$ **30.** $1\dfrac{2}{5}$ **31.** $2\dfrac{1}{25}$ **32.** $7\dfrac{49}{50}$

Write each percent as a decimal.

33. 51%   **34.** 52.3%   **35.** 102%   **36.** 2.5%

37. 3%   **38.** 0.1%   **39.** 9%   **40.** 234%

Write each percent as a fraction or mixed number in simplest form.

41. 99%   **42.** 125%   **43.** 50%   **44.** 3%

45. 64%   **46.** 12%   **47.** 150%   **48.** 8%

49. 2%   **50.** 4%   **51.** 225%   **52.** 120%

Topic 2 Warm-ups

Standardized Testing Warm-Ups

1. Which decimal is equal to the fraction $\dfrac{3}{5}$?

    A 0.15   **B** 0.6   **C** 0.35   **D** 1.67

2. Which fraction is equal to the decimal 0.375?

    A $\dfrac{2}{3}$ **B** $\dfrac{3}{8}$ **C** $\dfrac{75}{2}$ **D** $\dfrac{3}{75}$

Homework Review Warm-ups

Write each percent as a decimal and as a fraction or mixed number in simplest form.

3. 135%   **4.** 7%   **5.** 16%   **6.** 28%

Write each fraction or mixed number as a percent. Round to the nearest tenth of a percent.

7. $\dfrac{4}{11}$ **8.** $\dfrac{13}{12}$ **9.** $6\dfrac{3}{20}$ **10.** $\dfrac{1}{9}$

Topic 2   2   *Rates and Ratios*

GOAL

Use ratios and rates to compare quantities. Find unit rates from words and graphs.

Recall from the previous lesson that in 1990, about 51 of every 100 people in the United States were females. Therefore, about 49 of every 100 people were males. The ratio of females to males was about $\frac{51}{49}$.

Terms to Know	*Example / Illustration*
Ratio of a to b ($b \neq 0$) the relationship $\frac{a}{b}$ of two quantities a and b ($b \neq 0$)	51 to 49, $\frac{51}{49}$, or 51:49
Rate of a per b ($b \neq 0$) the relationship $\frac{a}{b}$ of two quantities a and b ($b \neq 0$) that are measured in different units	A rate of 110 mi to 2 h indicates the ratio of distance traveled to travel time.
Unit rate a rate per one given unit	The unit rate for the rate above is $\frac{55 \text{ mi}}{1 \text{ h}}$ or 55 miles per hour.

UNDERSTANDING THE MAIN IDEAS

The ratio of one number a to another number b ($b \neq 0$) is the quotient when a is divided by b. There are three ways to express the ratio:

$$a \text{ to } b \qquad \frac{a}{b} \qquad a:b$$

To write a ratio in simplest form, simply write the ratio as a fraction in simplest form. However, do not change the fraction to a mixed number. If the simplified fraction has a denominator of 1, be sure to write it as the second value in the ratio.

Example 1

The table shows the records of four of the 1999 division champions in the National Football League. Write each ratio in simplest form.

Team	Wins	Losses
Jacksonville	14	2
St. Louis	13	3
Washington	10	6
Seattle	9	7

a. Jacksonville's wins to losses

b. Washington's wins to losses

c. St. Louis's losses to Washington's losses

■ Solution ■

a. $\dfrac{14}{2} = \dfrac{7}{1}$ 　　　　 b. $\dfrac{10}{6} = \dfrac{5}{3}$ 　　　　 c. $\dfrac{3}{6} = \dfrac{1}{2}$

Refer to the table in Example 1. Write each ratio in simplest form.

1. St. Louis's wins to losses 　　　　 **2.** Seattle's losses to St. Louis's losses

If a rate compares two like quantities measured in different units, first write the measurements so the units are the same.

Example 2

Write the ratio of 27 yd to 18 ft in simplest form.

■ Solution ■

Rewrite the measures so they have the same units. There are two methods: (1) rewrite in feet or (2) rewrite in yards.

Method 1: Change 27 yd to feet: since 1 yd = 3 ft, 27 yd → 27 · 3 = 81 → 81 ft

$$\frac{27 \text{ yd}}{18 \text{ ft}} = \frac{81 \text{ ft}}{18 \text{ ft}} = \frac{81}{18} = \frac{9}{2}$$

Method 2: Change 18 ft to yards: since 1 ft = $\frac{1}{3}$ yd, 18 ft → $18\left(\frac{1}{3}\right) = 6 \rightarrow 6$ yd

$$\frac{27 \text{ yd}}{18 \text{ ft}} = \frac{27 \text{ yd}}{6 \text{ yd}} = \frac{27}{6} = \frac{9}{2}$$

Notice that both choices result in the same ratio.

Write each ratio as a fraction in simplest form.

3. 5 km to 300 m   **4.** 3 weeks : 8 days   **5.** 12 ft to 8 in.

A *unit rate* compares a quantity to a single unit of a different quantity. For example, the speed of a car might be 55 miles per hour. This is the unit rate $\dfrac{55 \text{ mi}}{1 \text{ h}}$. Notice that the bottom quantity is 1, and that the units (miles and hours) are different.

Example 3

Write the unit rate.

a. $48 in 3 h

b. 180 mi in 3 h

■ Solution ■

a. Use the rate $\dfrac{\text{dollars}}{\text{hours}}$: $\dfrac{48}{3} = \dfrac{16}{1}$.

The unit rate is $\dfrac{\$16}{1 \text{ h}}$, or

$16 per hour ($16/h).

b. Use the rate $\dfrac{\text{miles}}{\text{hours}}$: $\dfrac{180}{3} = \dfrac{60}{1}$.

The unit rate is $\dfrac{60 \text{ mi}}{1 \text{ h}}$, or

60 miles per hour (60 mi/h).

Write the unit rate.

6. $10 for 2 lb

7. 1200 mi in 4 days

8. $1160 in 4 weeks

9. 52 m in 4 s

· · · · · · · · · · · · · · · · · · ·
Spiral Review

10. Write 86% as a decimal and as a fraction in simplest form.

11. Write $2\dfrac{9}{10}$ as a decimal and as a percent.

Topic 2  Practice

The table below shows the approximate number of men and women on active duty in the United States armed forces in 1998.

Branch	Number of women (thousands)	Number of men (thousands)
Army	72	412
Air Force	66	301
Navy	51	332
Marine Corps	10	163

Use the table to write each ratio in simplest form.

1. women to men in the Army

2. men to women in the Marine Corps

3. women in the Navy to women in the Army

4. men in the Navy to men in the Army

5. men in the Army to men in all four branches of the armed services

Write each ratio in simplest form.

6. 63 to 35   **7.** 120 to 80   **8.** 28 ft : 6 in.

9. 6 lb to 4 oz   **10.** 10 m to 2 cm   **11.** 3 days to 8 h

12. 3 h : 15 min   **13.** $12 to $2.50   **14.** 2 days to 3 h

Write the unit rate.

15. 270 mi in 6 h   **16.** $51 in 6 h

17. 16 lb in 8 weeks   **18.** $30 for 4 tickets

19. $8.25 for 3 lb   **20.** 300 words in 5 min

21. 15 m in 2 s   **22.** $1800 in 12 months

23. $21 for 15 gal   **24.** 180 km in 2 h

25. 135 revolutions in 3 min   **26.** 200 students to 8 teachers

Topic 2 ③ *Warm-ups*

Standardized Testing Warm-ups

1. Which of the following correctly expresses the ratio 8 hours : 7 days as a fraction in simplest form?

   A $\dfrac{8}{7}$ B $\dfrac{8}{168}$ C $\dfrac{1}{21}$ D $\dfrac{1}{7}$

2. You are a sales representative for a sporting goods company. You make 30 sales calls and sell $2400 of merchandise. Estimate the amount of sales, in dollars per sales call.

   A $8 per sales call   B $80 per sales call

   C $125 per sales call   D $800 per sales call

Homework Review Warm-ups

In Exercises 3–5, write each ratio in simplest form.

3. 165 apples : 135 apples

4. 10 gallons to 15 gallons

5. 40 hours to 35 hours

Topic 2 **3** *Applying Percent*

GOAL
Use percents.

Most states have a sales tax, expressed as a percent. For example, if the sales tax rate is 6%, this means that for each dollar you pay, you must pay an extra six cents in tax. To find how much tax an item has, you need to find 6% of the cost of the item.

UNDERSTANDING THE MAIN IDEAS

There are two methods to find a percent of a given number: (1) write the percent as a decimal, then multiply, or (2) write the percent as a fraction, then multiply.

Example 1

What is 15% of 60?

Solution

Method 1: $15\% = 0.15$ Write 15% as a decimal.

$0.15 \times 60 = 9$ Multiply.

Method 2: $15\% = \dfrac{15}{100} = \dfrac{3}{20}$ Write 15% as a fraction.

$\dfrac{3}{20} \times \dfrac{60}{1} = \dfrac{180}{20} = 9$ Multiply. Write in simplest form.

15% of 60 is 9.

Find the percent of the number.

1. 50% of 274

2. 75% of 836

3. 14% of 1203

4. 9% of 523.6

5. 0.5% of 123

6. 106% of 95.6

You can solve many real-life problems by finding a percent of a number.

Example 2

The regular price of a video game is $54.90. It is on sale for 30% off the regular price. What is the discount? What is the sale price?

▪ Solution ▪

Discount = Percent off × Regular Price

= 30% × 54.90	Substitute.
= 0.3 × 54.90	Rewrite 30% as 0.3.
= $16.47	Multiply.

The discount is $16.47. To find the sale price, subtract the discount from the regular price. So, $54.90 − $16.47 = $38.43. The sale price is $38.43.

7. The regular price of a VCR is $179.90. It is on sale for 15% off the regular price. How much is the discount? What is the sale price of the VCR?

8. The sales tax on the VCR in Exercise 7 is 5.25%. What is the amount of sales tax on the discounted VCR? What is the total sale price of the VCR?

To express a comparison as a percent, first write the comparison in fraction form. Then write the fraction as a percent.

Example 3

6 out of 25 is what percent?

▪ Solution ▪

Method 1: Divide to find the percent.

$$\frac{6}{25} = 6 \div 25 = 0.24 = 24\%$$

Method 2: Use equivalent fractions to find the percent.

$$\frac{6}{25} = \frac{6 \cdot 4}{25 \cdot 4} = \frac{24}{100} = 24\%$$

6 out of 25 is 24%

Find the percent. If necessary, round to the nearest tenth of a percent.

9. 7 out of 20   **10.** 5 out of 16   **11.** 9 out of 50   **12.** 12 out of 125

13. a. Explain why Exercise 11 is easy to do using equivalent fractions.

 b. Explain why Exercise 12 is easier to do using division than equivalent fractions.

Example 4

In a survey of 500 people, 240 named Italian as one of their favorite types of food. Find the percent of people who named Italian as one of their favorite types of food.

■ Solution ■

$$\frac{240}{500} = \frac{240 \div 5}{500 \div 5} = \frac{48}{100} = 48\%$$

48% of the people surveyed named Italian as one of their favorite types of food.

Exercises 14–16 give the number of households with a type of pet out of 80,000 households surveyed. Find the percent of households with each type of pet. If necessary round to the nearest tenth of a percent.

14. Dog: 25,600   **15.** Cat: 21,600   **16.** Bird: 3680

· · · · · · · · · · · · · · · · · · · ·
Spiral Review

17. Write the ratio of 1 inch to 20 feet in simplest form.

18. Write the fraction $\frac{24}{25}$ as a decimal and as a percent.

Topic 2 **3** *Practice*

In Exercises 1–12, find the percent of the number.

1. 30% of 80   **2.** 20% of 30   **3.** 15% of 240   **4.** 85% of 300

5. 55% of 125   **6.** 120% of 42   **7.** 7% of 28   **8.** 47% of 356

9. 165% of 684   **10.** 0.5% of 19   **11.** 3% of 6574   **12.** 112% of 87.6

In Exercises 13–21, find the percent. If necessary, round to the nearest tenth of a percent.

13. 4 out of 5   **14.** 1 out of 8   **15.** 9 out of 40

16. 5 out of 6   **17.** 5 out of 9   **18.** 7 out of 30

19. 64 out of 100   **20.** 45 out of 180   **21.** 38 out of 52

22. If the sales tax rate is 6.5%, how much tax will you pay for an item whose price is $20?

23. The regular price of a book is $16.40. A store is having a 20% off sale. How much is the discount? What is the sale price?

24. You have a collection of 123 books. If 49 of these books are mysteries, what percent of the books in your collection are mysteries? Round to the nearest whole percent.

25. You work in a restaurant for $5.75 per hour. You are given an 8% raise. By how much does your hourly wage increase? What is your new hourly wage?

In Exercises 26–28, use the table at the right. It shows the percent of urban land area and rural land area out of the total land area for different states. For each state, how many square miles are urban land? How many square miles are rural land?

State	Urban	Rural
California	5.2%	94.8%
New Jersey	32.7%	67.3%
Nevada	0.9%	99.1%

26. California   155,973.2 square miles

27. New Jersey   7,418.8 square miles

28. Nevada   109,805.5 square miles

Topic 3   *Warm-ups*

Standardized Testing Warm-ups

1. The cost of Minh's restaurant meal is $9.75. If Minh plans to tip the server 20% of that amount, which is the best estimate of the tip?

   A $1.00

   B $1.25

   C $1.50

   D $2.00

2. Out of 27 students in a class, 7 chose purple as their favorite color. What percent of the class chose purple?

   A about 20%

   B about 26%

   C about 2.6%

   D about 3.9%

Homework Review Warm-ups

3. 42 out of 112 is what percent?

4. 0.6% of 225 is what number?

5. An advertisement claimed that 80% of the dentists surveyed preferred a certain toothpaste over every other leading brand. If only 15 dentists took part in the survey, how many preferred that toothpaste?

Topic 3 LESSON 1 *Adding Integers*

GOAL
Add integers.

Number lines not only show the relationships between positive and negative numbers, but they are very helpful in performing operations with positive and negative numbers.

Terms to Know	*Example / Illustration*
Integer any number that is a positive or negative whole number or zero (On a horizontal number line, the negative integers are to the left of 0 and the positive integers are to the right of 0.)	-11 41 0 23 -9
Opposites two numbers whose sum is zero	
Absolute value on a number line, the distance from the number to 0 (The symbol $|x|$ is read "the absolute value of x.")	$-2 + 2 = 0$, so -2 and 2 are opposites. $|2| = 2$ and $|-2| = 2$

UNDERSTANDING THE MAIN IDEAS

You can draw arrows on a number line to model the addition of two integers. As shown in the following example, an arrow pointing to the left, the negative direction, on the number line models adding a negative number. An arrow pointing to the right, the positive direction, models adding a positive number.

Example 1

Use a number line to find each sum.

 a. $-1 + (-3)$ **b.** $6 + (-3)$

▪ Solution ▪

a.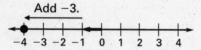

Start at 0 and draw an arrow 1 unit to the left to -1. From the tip of this arrow, draw a second arrow 3 more units to the left to add -3. The tip of this second arrow models the sum: $-1 + (-3) = -4$.

b.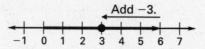

Start at 0 and draw an arrow 6 units to the right to 6. From the tip of this arrow, draw a second arrow 3 units to the left to add -3. The tip of this second arrow models the sum: $6 + (-3) = 3$.

Use a number line to find each sum.

1. $2 + 3$ **2.** $-3 + (-2)$ **3.** $-4 + (-2)$

When adding two integers with the same sign, follow these steps:

1. Find the absolute value of each number.

2. Add the two absolute values.

3. Give the result the same sign as the two numbers.

Example 2

Find the sum $-7 + (-9)$.

▪ Solution ▪

Step 1: Find the two absolute values: $|-7| = 7$; $|-9| = 9$.

Step 2: Add the two absolute values: $7 + 9 = 16$.

Step 3: Give the result the same sign as the numbers: $-7 + (-9) = -16$.

Find each sum.

4. $11 + 19$ **5.** $-13 + (-20)$ **6.** $-9 + (-11)$

When adding two integers with different signs, follow these steps:

1. Find the absolute value of each number.

2. Subtract the lesser absolute value from the greater absolute value.

3. Give the result the same sign as the number with the greater absolute value.

Example 3

Find the sum $-10 + 4$.

■ Solution ■

Step 1: Find the two absolute values: $|-10| = 10$; $|4| = 4$.

Step 2: Subtract the lesser absolute value from the greater absolute value:
$10 - 4 = 6$.

Step 3: Give the result the same sign as the number with the greater absolute value:
$-10 + 4 = -6$.

Find each sum.

7. $-15 + 7$

8. $2 + (-11)$

9. $7 + (-13)$

....
Spiral Review

10. What is the simplest form of the quotient $\dfrac{6}{11} \div \dfrac{7}{22}$?

11. What percent of 200 is 170?

Topic 3   *Practice*

Find each absolute value.

1. $|21|$ **2.** $|-11|$ **3.** $|34|$

4. $|-34|$ **5.** $|15|$ **6.** $|-4|$

7. $|-19|$ **8.** $|0|$ **9.** $|13|$

For Exercises 10–42, find each sum.

10. $-5 + (-5)$ **11.** $4 + 6$ **12.** $-5 + 3$

13. $7 + (-4)$ **14.** $-3 + 2$ **15.** $6 + (-7)$

16. $-43 + 51$ **17.** $17 + (-15)$ **18.** $-98 + 16$

19. $44 + (-61)$ **20.** $22 + 65$ **21.** $-19 + (-33)$

22. $88 + 12$ **23.** $50 + (-25)$ **24.** $-70 + (-8)$

25. $40 + (-29)$ **26.** $-12 + (-52)$ **27.** $-18 + 25$

28. $37 + (-19)$ **29.** $-29 + (-16)$ **30.** $71 + 54$

31. $-32 + 32$ **32.** $-40 + 88$ **33.** $95 + (-55)$

34. $18 + 87$ **35.** $-100 + 81$ **36.** $51 + 17$

37. $15 + (-9)$ **38.** $-21 + (-60)$ **39.** $76 + (-76)$

40. $-18 + 34$ **41.** $-31 + (-22)$ **42.** $75 + 29$

43. At 8 a.m., the temperature was $-5°F$. Over the next five hours, the temperature rose $16°F$. What was the temperature at 1 P.M.?

44. A football team lost 15 yards on one play. On the next play, they gained 8 yards. Find the total amount of yardage gained or lost on these two plays.

Topic 3   *Warm-ups*

Standardized Testing Warm-ups

1. Find the sum: $-10 + 12$

  A 2   **B** -2 **C** 22   **D** -22

2. Find the sum: $(-3) + (-4)$

  A -1 **B** 1   **C** -7 **D** 7

Homework Review Warm-ups

Find the sum.

3. $-27 + 13$ **4.** $27 + (-13)$ **5.** $-13 + (-27)$

Topic 3 LESSON 2 — *Subtracting Integers*

GOAL

Subtract integers.

At 6:00 A.M. on the morning of January 12, 1911, the temperature in Rapid City, South Dakota was 49°F. Over the next two hours, the temperature dropped 62°F! At 8:00 A.M, the temperature was −13°F. The temperature change can be represented by subtracting integers:

$$49 - 62 = -13.$$

UNDERSTANDING THE MAIN IDEAS

When you subtract two integers, you add the opposite of the integer you are subtracting: $a - b = a + (-b)$

You can draw arrows on a number line to model the subtraction of integers. When you subtract, draw an arrow in the "opposite" direction:

To subtract a positive number, draw an arrow pointing to the left, the negative direction.

To subtract a negative number, draw an arrow pointing to the right, the positive direction.

The examples will show you both methods for subtracting integers. Example 1 and Example 2 show how to subtract a positive integer.

Example 1

Find the difference $2 - 5$.

■ Solution ■

Method 1: Add the opposite.

$2 - 5 = 2 + (-5)$

$\qquad = -3$

Method 2: Use a number line.

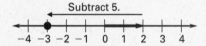
Subtract 5.

Start at 0. Draw an arrow 2 units to the right. From the tip of the arrow, draw an arrow 5 units to the left to subtract 5. The tip of the arrow models the difference: $2 - 5 = -3$

Example 2

Find the difference $-2 - 4$.

■ Solution ■

Method 1: Add the opposite.

$-2 - 4 = -2 + (-4)$

$\qquad = -6$

Method 2: Use a number line.

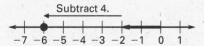

Start at 0. Draw an arrow 2 units to the left to -2. From the tip of the arrow, draw an arrow 4 units to the left to subtract 4. The tip of the arrow models the difference: $-2 - 4 = -6$

Find each difference.

1. $5 - 9$ **2.** $7 - 5$ **3.** $-14 - 12$ **4.** $-10 - 3$

Example 3 and Example 4 show how to subtract a negative integer. Remember that the opposite of a negative is a positive. For example, $-(-8) = 8$.

Example 3

Find the difference $3 - (-5)$.

■ Solution ■

Method 1: Add the opposite.

$3 - (-5) = 3 + [-(-5)]$

$\qquad = 3 + 5$

$\qquad = 8$

The opposite of -5 is 5.

Method 2: Use a number line.

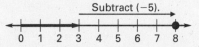

Start at 0. Draw an arrow 3 units to the right. From the tip of the arrow, draw an arrow 5 units to the right to subtract -5. The tip of the arrow models the difference: $3 - (-5) = 8$

Example 4

Find the difference $-1 - (-4)$.

■ Solution ■

Method 1: Add the opposite.

$$-1 - (-4) = -1 + [-(-4)]$$
$$= -1 + 4$$
$$= 3$$

The opposite of -4 is 4.

Method 2: Use a number line.

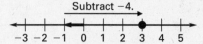
Subtract -4.

Start at 0. Draw an arrow 1 unit to the left to -1. From the tip of the arrow, draw an arrow 4 units to the right to subtract -4. The tip of the arrow models the difference: $-1 - (-4) = 3$

Give the opposite of each number.

5. 5   **6.** -4 **7.** -12 **8.** 32

Find each difference.

9. $9 - (-14)$ **10.** $27 - (-3)$ **11.** $-16 - (-4)$ **12.** $-7 - (-8)$

· · · · · · · · · · · · · · · · · ·
Spiral Review

Find each sum.

13. $11 + (-9)$ **14.** $7 + (-4)$ **15.** $-12 + 14$ **16.** $-10 + 3$

Topic 3   *Multiplying and Dividing Integers*

GOAL

Multiply and divide integers.

When you add or subtract integers, the sign of the result depends on the values of the integers. When you multiply or divide integers, the sign of the result depends only on the signs of the two integers.

UNDERSTANDING THE MAIN IDEAS

In the following example, chips are used to show that the product of a negative integer and a positive integer is negative.

Example 1

Find the product $4(-3)$.

■ Solution ■

Think of the product as 4 times a group of 3 negative chips, that is, 4 groups of 3 negative chips. When these groups are combined, there are 12 negative chips representing the product -12.

4 groups of 3   1 group of 12
negative chips   negative chips

 →

$$4(-3) \quad = \quad -12$$

Find each product.

1. $5(-2)$ **2.** $3(-3)$ **3.** $2(-4)$

Since the order in which two numbers are multiplied does not affect the product, $4(-3) = (-3)(4)$. This suggests that the product of two integers with different signs is always negative.

When you multiply two positive integers, you know that the product is positive. But what happens when you multiply two negative integers?

Example 2

a. Find the products $3(-3)$, $2(-3)$, $1(-3)$, and $0(-3)$.

b. Identify the pattern in the products found in part (a) and use it to find the products $(-1)(-3)$, $(-2)(-3)$, and $(-3)(-3)$.

Solution

a. Use the method of Example 1. Think of the products as 3 times a group of 3 negative chips, 2 times a group of 3 negative chips, 1 times a group of 3 negative chips, and 0 times a group of 3 negative chips, respectively. Therefore, the products are:

$3(-3) = -9$, $2(-3) = -6$, $1(-3) = -3$, and $0(-3) = 0$.

b. In part (a), each time the number multiplied by -3 decreases by 1, the product increases by 3. This pattern suggests that the products are:

$(-1)(-3) = 3$, $(-2)(-3) = 6$, and $(-3)(-3) = 9$.

In part (b) of Example 2, notice that when -3 is multiplied by a negative number, the result is positive. This suggests that when two integers have the same sign, their product is positive.

Multiplying Integers

The product of two integers with the same sign is positive.
The product of two integers with different signs is negative.

Example 3

Find the product $(-17)(30)$.

Solution

The two integers have different signs, so the product is negative. Thus, just multiply the two integers as if they were both positive and then attach a negative sign to the result: $(-17)(30) = -510$.

For Exercises 4–6, find each product.

4. $(-5)(12)$ **5.** $8(-4)$ **6.** $(-16)(6)$

7. Determine whether the product $3(-4)(-5)$ is positive or negative. Explain.

To divide two integers, recall the relationship between multiplication and division.

$12 \div 3 = 4$ because $4(3) = 12$. $-12 \div 3 = -4$ because $(-4)(3) = -12$.

$12 \div (-3) = -4$ because $(-4)(-3) = 12$. $-12 \div (-3) = 4$ because $4(-3) = 12$.

The rules for dividing two integers are based on those for multiplying two integers.

Dividing Integers

The quotient of two integers with the same sign is positive.

The quotient of two integers with different signs is negative.

Example 4

Find each quotient.

a. $36 \div (-9)$ **b.** $-36 \div (-9)$

Solution

a. The signs are different, so the quotient is negative. $36 \div (-9) = -4$

b. The signs are the same, so the quotient is positive. $-36 \div (-9) = 4$

Find each quotient.

8. $-28 \div (-7)$ **9.** $36 \div (-12)$ **10.** $-64 \div 16$

. . . .
Spiral Review

Write the unit rate.

11. $30 in 5 h   **12.** 150 mi in 4 h   **13.** 320 words in 5 min

Topic 3   *Practice*

For Exercises 1–39, find each product or quotient.

1. $12(-7)$ 2. $64 \div (-4)$ 3. $90 \div 5$

4. $-88 \div 11$ 5. $(-24)(10)$ 6. $98 \div (-7)$

7. $15(20)$ 8. $30 \div 6$ 9. $(-18)(-3)$

10. $-82 \div 2$ 11. $-105 \div (-7)$ 12. $400 \div (-20)$

13. $42(-3)$ 14. $(-11)(-11)$ 15. $-78 \div 6$

16. $26 \div (-2)$ 17. $(-2)(52)$ 18. $17(3)$

19. $22(-5)$ 20. $75 \div (-3)$ 21. $(-12)(-13)$

22. $(-8)(22)$ 23. $(-21)(-4)$ 24. $42 \div (-7)$

25. $-125 \div (-5)$ 26. $(-5)(-16)$ 27. $(-3)(33)$

28. $250 \div (-5)$ 29. $68 \div (-4)$ 30. $55 \div 5$

31. $(-25)(-40)$ 32. $-48 \div 16$ 33. $(-4)(-5)(-10)$

34. $96 \div (-4)$ 35. $(-4)(-4)(-4)$ 36. $-16 \div (-2)$

37. $(-32)(9)$ 38. $2(-12)(-3)$ 39. $56 \div (-7)$

40. A nurse checked a patient's temperature every hour for 3 h. Each time, the temperature had fallen 1°F. Express the change in the patient's temperature over that time period as an integer.

41. A football team lost 5 yd on each of three consecutive plays. Express the total yardage for the three plays as an integer.

Topic 3 | 4 | *Warm-ups*

PRE-COURSE REVIEW
LESSON

Standardized Testing Warm-ups

1. Find the product: $(25)(-5)$

    A -5 **B** 5 **C** -125 **D** 125

2. Find the quotient: $25 \div (-5)$

    A -5 **B** 5 **C** -125 **D** 125

Homework Review Warm-ups

Find each product or quotient.

3. $(-6)(-7)$ **4.** $-90 \div 15$ **5.** $10(-18)$ **6.** $-55 \div (-5)$

Topic 3 LESSON 4 *Order of Operations*

GOAL

Evaluate numerical expressions with several operations and grouping symbols.

Suppose you want to evaluate the expression $2 + 3 \cdot 5$. If you add first, you get $5 \cdot 5 = 25$. If you multiply first, you get $2 + 15 = 17$. To avoid confusion, there are rules that we use to evaluate an expression so that everyone gets the same answer.

Terms to Know | ## *Example / Illustration*

Terms to Know	Example / Illustration
Grouping symbols parentheses ( ), brackets [ ], absolute value symbol, and fraction bars that indicate the order in which the operations should be done	$2(1 - 5) = 2(-4) = -8$ $\dfrac{5 + 3}{2} = \dfrac{8}{2} = 4$
Numerical expression a collection of numbers, operations, and grouping symbols	$\dfrac{5 + 3}{2}$
Simplify a numerical expression to evaluate (that is, find the value of) the expression	$\dfrac{5 + 3}{2} = \dfrac{8}{2} = 4$
Power the result of repeated multiplication	$(-2)^3 = (-2)(-2)(-2) = -8$; $(-2)^3$ is read "negative two to the third power." -2 is the base. 3 is the exponent.
Order of operations rules that tell you in what order to do the operations to evaluate a numerical expression	$12 \div (4 - 2) + 3^2 - 5 \times 2$ $= 12 \div 2 + 3^2 - 5 \times 2$ $= 12 \div 2 + 9 - 5 \times 2$ $= 6 + 9 - 10$ $= 5$

UNDERSTANDING THE MAIN IDEAS

When you evaluate a numerical expression that involves more than one operation, you should use the order of operations that everyone agrees on so that you get the same results as everyone else. This is the order you should use:

1. First do operations that occur within grouping symbols.

2. Then evaluate powers.

3. Then do multiplications and divisions from left to right.

4. Finally, do additions and subtractions from left to right.

Keep in mind that multiplication can be written in various ways. For example, $4 \cdot 2$, $4(2)$, $(4)(2)$, and 4×2 all mean "4 times 2." Division can also be written in more than one way. For example, $14 \div 2$ and $\frac{14}{2}$ both mean "14 divided by 2."

Example

Evaluate each expression.

a. $\frac{1}{2} \cdot 4^3 - 18 \div 6 + 1$

b. $\left| 15 - 2^4 \right|$

c. $\frac{4 \cdot 3 + 8}{9 - 4}$

d. $\frac{8}{16} - 7(3 - 5) - 16$

Solution

a. There are no grouping symbols, so evaluate the power first.

$\frac{1}{2} \cdot 4^3 - 18 \div 6 + 1$ $\leftarrow$ Evaluate the power 4^3: $4^3 = 4 \cdot 4 \cdot 4 = 64$.

$= \frac{1}{2} \cdot 64 - 18 \div 6 + 1$ $\leftarrow$ Then do the multiplication and division from left to right.

$= 32 - 3 + 1$ $\leftarrow$ Next do the addition and subtraction from left to right.

$= 30$

b. Evaluate the expression inside the absolute value bars first. Start by evaluating the power.

$\left| 15 - 2^4 \right| = \left| 15 - 16 \right| = \left| -1 \right| = 1$

Remember: The absolute value of a negative number is its opposite.

■ **Solution** ■ *continued*

c. A fraction bar is a grouping symbol. To evaluate an expression with a fraction bar, you must evaluate the numerator and the denominator before dividing. Follow the order of operations to simplify the numerator and the denominator.

$$\frac{4 \cdot 3 + 8}{9 - 4} \qquad \leftarrow \text{Do the multiplication first.}$$

$$= \frac{12 + 8}{9 - 4} \qquad \leftarrow \text{Then add and subtract.}$$

$$= \frac{20}{5} \qquad \leftarrow \text{Now you can divide.}$$

$$= 4$$

d. $\dfrac{18}{6} - 7(3 - 5) - 16 \quad \leftarrow$ First do the subtraction inside the parentheses.

$$= \frac{18}{6} - 7(-2) - 16 \quad \leftarrow \text{Then do the multiplication and division from left to right.}$$

$$= 3 - (-14) - 16 \quad \leftarrow \text{Next do the subtractions from left to right.}$$
$$\qquad\qquad\qquad\qquad \textit{Remember: } 3 - (-14) = 3 + 14$$

$$= 1$$

Evaluate each expression.

1. $7 + 2 \cdot 3$ 2. $(7 + 2) \cdot 3$ 3. $|18 \div 3 - 6|$

4. $7 - 4 \cdot 8$ 5. $(7 - 4) \cdot 8$ 6. $7 - (-4 \cdot 8)$

7. $2 - 5 \cdot 3^2$ 8. $2 - (5 \cdot 3)^2$ 9. $(2 - 5) \cdot 3^2$

10. $\dfrac{6 + 2}{6 - 2}$ 11. $\dfrac{7 - 2^2}{2 \cdot 3}$ 12. $\dfrac{6 + 1}{2 + 4 \cdot 3}$

• • • • • • • • • • • • • • • • • • • •
Spiral Review

13. What is 25% of 80?   14. Find the quotient: $-12 \div 4$

15. Find the product: -12×4 16. Write 85% as a fraction.

Topic 3  *Practice*

Evaluate each expression.

1. $7 - 3 + 1$

2. $5 \cdot 8 \div 2$

3. $5 \cdot 8 - 2$

4. $6 - 3(-3)$

5. $3(-4) - 8 \div 2$

6. $4 + (-8)(-2) - 9$

7. $|18 \div (3 - 6)|$

8. $1 + 2^3$

9. $(1 + 2)^3$

10. $5 + 2^2 - 3(6)$

11. $3^2 - 2^4 \div 4$

12. $8(3 + 1) \div 4^2$

13. $\dfrac{3 + 3}{2(5) + 2}$

14. $\dfrac{3^2 - 1}{2^3}$

15. $(7 - 3)2 + \dfrac{18}{6}$

16. $(2 + 3)^3 - (-2)^5$

17. $(1 + 3)^2 \div 2^3$

18. $-8 \cdot 2 + 6^2 \div 9$

19. $(5 - 2)(7 + 1)$

20. $4(2 + 32)$

21. $12 \div 3 + 8 \div 4$

22. $\dfrac{1 + 3(5)}{11 - 3(3)}$

23. $\dfrac{(1 + 3)(5)}{(11 - 3)(3)}$

24. $\dfrac{1 + 3(5)}{(11 - 3)(3)}$

25. $|9 - 4^3|$

26. $|(-3)^2 + (-6)|$

27. $9 + |5(-3)|$

For Exercises 28–31, two calculators were used to evaluate the expression. They gave different results. Determine which calculator performed the correct order of operations.

28. $4 \boxed{+} 6 \boxed{\div} 2 \boxed{-} 3 \boxed{=}$

Calculator A gave the answer 2. Calculator B gave the answer 4.

29. $6 \boxed{-} 6 \boxed{\times} 2 \boxed{\div} 3 \boxed{=}$

Calculator A gave the answer 2. Calculator B gave the answer 0.

30. $2 \boxed{\times} 3 \boxed{-} 5 \boxed{\times} 4 \boxed{=}$

Calculator A gave the answer 4. Calculator B gave the answer -14.

31. $12 \boxed{-} 6 \boxed{\div} 2 \boxed{\times} 3 \boxed{=}$

Calculator A gave the answer 3. Calculator B gave the answer 9.

Topic 3   **5**   *Warm-ups*

Standardized Testing Warm-ups

1. Evaluate the expression $\dfrac{6 + 2}{3 + 1} - 1$.

    A 1   **B** $1\dfrac{3}{4}$ **C** 2   **D** 3

2. Evaluate the expression $\dfrac{15 + 3 \div 3}{2} + 6 \times 2$.

    A 14   **B** 15   **C** 18   **D** 20

Homework Review Warm-ups

Evaluate each expression.

3. $6 - 4 \div 2$ **4.** $\dfrac{6 \times 7 - 2}{4}$ **5.** $\dfrac{8 - 4}{8 + 4}$ **6.** $15 - 3^2$

Topic 3 — LESSON 5 — *Evaluating Expressions*

GOAL

Evaluate variable expressions

In Lesson 4, you learned to evaluate numerical expressions. We use variables to represent numbers in variable expressions. For example, if you can buy any number of boxes of envelopes and each box contains 40 envelopes, then the expression $40n$ represents the number of envelopes in n boxes. To evaluate a variable expression such as $40n$, you substitute a reasonable value for n and then simplify the resulting numerical expression.

Terms to Know	*Example / Illustration*
Variable a letter that is used to represent one or more numbers	x a n
Value of a variable one or more numbers that a variable represents	If x is a whole number, then x can represent 0, 1, 2, 3, and so on.
Variable expression a collection of numbers, variables, operations, and grouping symbols	$\dfrac{ab}{2} \leftarrow a$ times b, divided by 2 $\dfrac{x}{3} + y \leftarrow x$ divided by 3, plus y
Value of a variable expression the number that results when each variable in a variable expression is replaced by a number and the expression is evaluated	If $x = 6$, then $2x + 3 =$ $2(6) + 3 = 12 + 3 = 15$.

UNDERSTANDING THE MAIN IDEAS

To evaluate a variable expression, you should follow these steps:

1. Substitute a value for each variable.

2. Use the order of operations to evaluate the resulting numerical expression.

The value of a variable expression depends on the values that are substituted for the variables.

Example 1

Evaluate the expression $x^2 - 2x + 5$ when $x = 3$.

■ Solution ■

Substitute 3 for x both times it appears in the expression. Even though you do not need a multiplication symbol to show the product of 2 and x, or $2x$, you must remember to include a multiplication symbol when you write the product of two numbers, such as 2 and 3.

$$x^2 - 2x + 5 = 3^2 - 2 \cdot 3 + 5 \quad \leftarrow \text{Substitute 3 for } x \text{ twice.}$$
$$= 9 - 2 \cdot 3 + 5 \quad \leftarrow \text{Simplify the power } 3^2 \text{ first.}$$
$$= 9 - 6 + 5 \quad \leftarrow \text{Simplify the product } 2 \cdot 3 \text{ next.}$$
$$= 8 \quad \leftarrow \text{Add and subtract, working in order from left to right.}$$

Evaluate each expression when (a) $x = 3$ and (b) $x = -2$.

1. $x - 2$　　　　　**2.** $6x$　　　　　**3.** $4x - 3$

4. $4x^2$　　　　　**5.** $5x - (2 + x)$　　　　　**6.** $2x^2 + 3x - 1$

A *formula* can be used to represent a relationship in a real-life situation.

Area of a rectangle	$A = \ell w$	A = area, ℓ = length, w = width
Temperature	$F = \dfrac{9}{5}C + 32$	F = degrees Fahrenheit, C = degrees Celsius
Distance traveled	$d = rt$	d = distance, r = rate or speed, t = time

Example 2

A car traveled at an average speed of 50 mi/h for 1.5 h. How many miles did it travel?

■ Solution ■

Use the formula $d = rt$, with $r = 50$ and $t = 1.5$.

$d = rt$ ← Write the distance formula.

$d = 50 \cdot 1.5$ ← Substitute 50 for r and 1.5 for t.

$d = 75$

The car traveled 75 miles.

---

7. A tennis court is 78 ft long and 36 ft wide. Find its area.

8. One side of a rectangular garden is 17 ft. The other is 15 ft. What is the area of the garden?

9. A bus traveled for $1\frac{1}{2}$ h at an average speed of 48 mi/h. How far did the bus travel?

10. Andy planned to walk 3.1 mi/h for 3 hours. How far did he plan to walk?

11. Suppose the Celsius temperature is 35°C. Find the temperature in degrees Fahrenheit.

· · · · · · · · · · · · · · · · · · · ·
Spiral Review

Evaluate each expression.

12. $-15 \div 3 + 7 \cdot 5$ **13.** $(-8)^2 - 6 \cdot 2 + 4$ **14.** $9(5 + 2) + 3$

15. $1 - 2^5 \div 4$ **16.** $(2 + 5 \cdot 8) \div 7$ **17.** $-25 - 3(-8) + 6$

Topic 3 **5** *Practice*

For Exercises 1–14, evaluate each expression.

1. $a - 17$ when $a = 21$ **2.** $12b$ when $b = -2$

3. $6.3x$ when $x = 100$ **4.** $1 + 2w$ when $w = 4.5$

5. $|x \div 6|$ when $x = 60$ **6.** $3t - 8$ when $t = -11$

7. $4a - (a - 5)$ when $a = 8$ **8.** $4x(2x - 1)$ when $x = 3$

9. $2(15 - c)$ when $c = -5$ **10.** $3(2 - 5n)$ when $n = 2$

11. $81 \div y + 7$ when $y = 3$ **12.** $x^2 + x - 6$ when $x = -3$

13. $\dfrac{3m + 2}{m - 1}$ when $m = 11$ **14.** $\left(\dfrac{y}{2}\right)^2 + 4y$ when $y = -6$

15. The Rosenthals drove 2.25 h at an average speed of 72 km/h. Use the formula $d = rt$ to find how many kilometers they traveled.

16. Find the area of a square floor tile if each side is 4 inches long. Use $A = s^2$.

17. Use the formula $F = \dfrac{9}{5}C + 32$ to find the temperature in degrees Fahrenheit when the temperature is 0°C.

18. An isosceles triangle has two sides of the same length. Write an expression for the perimeter of the isosceles triangle shown in the diagram. Use the expression to find the perimeter when $a = 5.5$ and $b = 4.5$.

Standardized Testing Warm-Ups

1. Evaluate the expression $5(x - 3)$ when $x = 7$.

   A 32 **B** -7 **C** 20 **D** 30

2. Evaluate expression $\dfrac{7a + 6}{a}$ when $a = -1.5$.

   A 2 **B** 3 **C** 4 **D** 5

Homework Review Warm-Ups

Evaluate each expression.

3. $\dfrac{z + 7}{z - 3}$ when $z = 5$

4. $\dfrac{r^2}{6} - 2$ when $r = 6$

5. $|3t - 5|$ when $t = -4$

6. $7y(y + 4)$ when $y = 1$

Topic 4  Conversions in the Customary System

GOAL

Convert customary system units for length, distance, weight, and capacity.

A soup recipe calls for 2 pounds of tomatoes. Tomatoes are sold in 8-ounce cans. To know how many cans of tomatoes to buy, you need to know how to convert between ounces and pounds.

Terms to Know

Terms to Know	Example / Illustration
Inch (in.), Foot (ft), Yard (yd), Mile (mi) customary system units for measuring length or distance	12 in. = 1 ft 3 ft = 1 yd 1760 yd = 1 mi Also, 36 in. = 1 yd 5280 ft = 1 mi
Ounce (oz), Pound (lb), Ton (T) customary system units for measuring weight	16 oz = 1 lb 2000 lb = 1 T
Fluid ounce (fl oz), Cup (c), Pint (pt), Quart (qt), Gallon (gal) customary system units for measuring capacity	8 fl oz = 1 c 2 c = 1 pt 2 pt = 1 qt 4 qt = 1 gal

UNDERSTANDING THE MAIN IDEAS

Using your number sense will help you convert between customary system units. To convert from a smaller unit, like inches, to a larger unit, like feet, divide. This makes sense because it takes fewer larger units than smaller units to describe a quantity. For example, a person who is 60 *inches* tall is 5 *feet* tall.

Example 1

The drama club is making 16 costumes for a school play. Each costume takes $2\frac{1}{2}$ feet of ribbon. The ribbon is sold in yards. At least how many yards of ribbon does the drama club need to buy?

▬ Solution ▬

Step 1: To find the total amount of ribbon needed in feet for all the costumes, multiply.

$$16 \times 2\frac{1}{2} \text{ ft} = 40 \text{ ft}$$

Step 2: You are converting from a smaller unit, feet, to a larger unit, yards, so *divide*.

$3 \text{ ft} = 1 \text{ yd}$ Divide by 3 to convert from feet to yards.

$$40 \text{ ft} \rightarrow 40 \div 3 = 13\frac{1}{3} \rightarrow 13\frac{1}{3} \text{ yd}$$

The drama club needs to buy at least $13\frac{1}{3}$ yards of ribbon.

Complete.

1. 18 ft = __?__ yd
2. 18 in. = __?__ ft
3. 108 in. = __?__ yd
4. 16 fl oz = __?__ c
5. 10 c = __?__ pt
6. 6 pt = __?__ qt
7. 20 qt = __?__ gal
8. 32 oz = __?__ lb
9. 4000 lb = __?__ T

To convert from a larger unit, like tons, to a smaller unit, like pounds, multiply. This makes sense because it takes more smaller units than larger units to describe a quantity. For example, a bull might weigh 1 *ton*, but it weighs 2000 *pounds*.

You may need to multiply or divide more than once to convert between customary system measurements.

Example 2

Kahlil bought four gallons of lemonade to serve at the business club meeting. How many one-cup servings can he serve if all of the lemonade is used?

■ **Solution** ■

You aren't given an equivalence between cups and gallons. You will have to perform more than one conversion.

Step 1: Find the number of cups in each gallon. Since you are converting from a larger unit, gallons, to a smaller unit, cups, you will need to *multiply*.

> 1 gallon = 4 quarts
>
> 4 quarts → 4 × 2 = 8 → 8 pints
>
> 8 pints → 8 × 2 = 16 → 16 cups
>
> So, 1 gallon = 16 cups.

Step 2: To find the number of cups of lemonade in 4 gallons, multiply the number of cups in 1 gallon by 4.

> 4 gallons → 4 × 16 = 64 → 64 cups

Four gallons of lemonade will make 64 one-cup servings.

Determine whether there will be more or fewer of the new units after the conversion. Then find the equivalent measure.

10. 2 c = __?__ fl oz   **11.** 3 pt = __?__ c   **12.** 3 T = __?__ lb

13. 10 lb = __?__ oz   **14.** 12 oz = __?__ lb   **15.** 1000 lb = __?__ T

16. 1 gal = __?__ pt   **17.** 8 qt = __?__ gal   **18.** 2 qt = __?__ pt

19. Hannah ran a 440-yard dash in $1\frac{1}{2}$ minutes.

    a. What fractional part of a mile did Hannah run?

    b. If Hannah could run a mile at the same speed she ran the dash, how long would it take her to run the mile?

20. Paulo needs 3 pounds of mushrooms for a sauce recipe. Mushrooms are sold in 4-ounce cans. What does Paulo need to know in order to buy the mushrooms? How many cans of mushrooms does Paulo need to make the recipe?

· · · · · · · · · · · · · · · ·
Spiral Review

Write each percent as a decimal.

21. 20%   **22.** 101%   **23.** 2%   **24.** 257%

Topic 4   *Practice*

Complete.

1. 6 yd = _?_ ft

2. 2 yd = _?_ in.

3. 10 mi = _?_ yd

4. 24 ft = _?_ yd

5. 15,840 ft = _?_ mi

6. 100 yd = _?_ ft

7. 24 yd = _?_ ft

8. 27 ft = _?_ yd

9. 8 yd = _?_ in.

10. $\frac{1}{2}$ gal = _?_ c

11. 3 qt = _?_ c

12. 16 fl oz = _?_ pt

13. 32 c = _?_ qt

14. $\frac{1}{2}$ pt = _?_ fl oz

15. 16 fl oz = _?_ c

16. 10 c = _?_ pt

17. 8 pt = _?_ qt

18. 1 qt = _?_ gal

19. 3 lb = _?_ oz

20. 2.5 T = _?_ lb

21. 1.5 lb = _?_ oz

22. $\frac{1}{2}$ lb = _?_ oz

23. $\frac{1}{2}$ T = _?_ lb

24. 160 oz = _?_ lb

25. 1 mi = _?_ in.

26. 1 T = _?_ oz

27. 1344 fl oz = _?_ gal

Solve.

28. Amber and Jason have put 6 qt of water into their fish tank. It holds 4 gal. How many more quarts of water do they need to put into the tank to fill it?

29. Rita had a piece of wood $3\frac{1}{2}$ feet long. She sawed off a piece 9 in. long to make a sign. How long was the piece of wood that was left?

30. Francis combined 24 oz of dry cereal, 8 oz of nuts, $\frac{3}{4}$ lb of pretzels, and $\frac{1}{4}$ lb of sesame sticks to make a trail mix. What was the total weight of the mix?

31. Grace ran $\frac{3}{4}$ mile, Hank ran 1000 yd, and Shawna ran 2640 ft. Order their distances from least to greatest.

Topic 4   2   *Warm-ups*

Standardized Testing Warm-Ups

1. Which quantity is the greatest?

   A 5 pt   **B** 75 fl oz   **C** 2 qt   **D** 7 c

2. How many yards are equivalent to $4\frac{1}{2}$ miles?

   A 7040 yd   **B** 4500 yd   **C** 23,760 yd   **D** 7920 yd

Homework Review Warm-Ups

3. Jerome is 72 inches tall. How tall is Jerome in feet?

4. A 3-gallon jug is filled with 5 quarts of liquid. Is the jug full? If not, how many more quarts will it take to fill it?

5. A fruit fizz recipe calls for 3 pints of orange juice, 3 cups of pineapple juice, 6 ounces of lemon juice, and a quart of seltzer water. Order the amounts of the ingredients from least to greatest.

Topic 4 · *Conversions in the Metric System*

GOAL
Convert metric system units for length, distance, mass, and capacity

You fill a glass from a full 1 liter bottle of water. The bottle now contains 750 milliliters of water. How much water is in the glass? To find out, you'll need to know how to convert metric system units.

Terms to Know	*Example / Illustration*
Millimeter (mm), Centimeter (cm), Meter (m), Kilometer (km) metric system units for measuring length or distance	10 mm = 1 cm 100 cm = 1 m 1000 m = 1 km Also, 1000 mm = 1 m
Milligram (mg), Gram (g), Kilogram (kg) metric system units for measuring mass (For most purposes, you can think of weight and mass as the same thing.)	1000 mg = 1 g 1000 g = 1 kg
Milliliter (mL), Liter (L), Kiloliter (kL) metric system units for measuring capacity	1000 ml = 1 L 1000 L = 1 kL

UNDERSTANDING THE MAIN IDEAS

Like our place value system, the metric measurement system is based on powers of 10. Just as in the customary system, multiply to convert from a larger unit to a smaller unit. Since the metric system is based on powers of 10, your multiplication will be by numbers like 10, 100, or 1000.

Example 1

Nicole is training to try out for the track team. She runs 1.2 km on Monday, 3.5 km on Wednesday, and 2.8 km on Friday. Her coach suggests that she run 7000 meters per week. Has Nicole run enough this week?

■ Solution ■

Step 1: Find the total distance that Nicole has run in kilometers.

$$1.2 \text{ km} + 3.5 \text{ km} + 2.8 \text{ km} = 7.5 \text{ km}$$

Step 2: You are converting from a larger unit, kilometers, to a smaller unit, meters, so multiply.

1 km = 1000 m   Multiply by 1000 to convert
from kilometers to meters.

7.5 km → 7.5 × 1000 = 7500 → 7500 m

7500 m > 7000 m

Nicole has run more than 7000 m this week.

Complete.

1. 3 km = __?__ m   **2.** 6 m = __?__ mm   **3.** 2 g = __?__ mg

4. 4 L = __?__ mL   **5.** 2.4 km = __?__ m   **6.** 9.1 m = __?__ mm

7. 1.2 kg = __?__ g   **8.** 8.97 m = __?__ cm   **9.** 4.26 g = __?__ mg

10. 6.752 L = __?__ mL   **11.** 0.342 m = __?__ cm   **12.** 0.9 kg = __?__ g

Divide to convert from a smaller unit to a larger unit. Since the metric system is based on powers of 10, your division will be by numbers like 10, 100, or 1000.

Example 2

Martha's punch mix recipe contains 750 mL of lime sherbet, 750 mL of orange sherbet, and 1500 mL of lemonade. How many liters of punch does Martha's recipe make?

■ Solution ■

Step 1: Find the total amount of ingredients in milliliters.

750 mL + 750 mL + 1500 mL = 3000 mL

Step 2: You are converting from a smaller unit, milliliters, to a larger unit, liters, so divide.

1000 mL = 1 L   Divide by 1000 to convert
                     from milliliters to liters.

3000 mL → 3000 ÷ 1000 = 3 → 3 L

The punch mix recipe makes 3 liters of punch.

Determine whether there will be more or fewer of the new units after the conversion. Then find the equivalent measure.

13. 3000 mg = __?__ g   **14.** 1400 g = __?__ kg   **15.** 1000 L = __?__ kL

16. 2.3 m = __?__ mm   **17.** 1.89 mL = __?__ L   **18.** 890 cm = __?__ m

19. 78 g = __?__ mg   **20.** 90.3 m = __?__ km   **21.** 6.7 m = __?__ cm

....
Spiral Review

Find each sum or difference.

22. 18 + (−22)   **23.** −1 − 7   **24.** −5 + 3

25. −7 + (−5)   **26.** 15 + 8   **27.** 4 − (−3)

PRE-COURSE REVIEW

Complete.

1. 4 m = __?__ km
2. 1 g = __?__ mg
3. 2 g = __?__ kg
4. 4000 mL = __?__ L
5. 2.8 km = __?__ cm
6. 9 L = __?__ kL
7. 37 kg = __?__ g
8. 480 cm = __?__ mm
9. 674 cm = __?__ m
10. 6.589 g = __?__ mg
11. 5.726 m = __?__ mm
12. 0.432 m = __?__ cm
13. 842 g = __?__ kg
14. 93 mL = __?__ L
15. 873 kL = __?__ L
16. 2374 mL = __?__ L
17. 375 mL = __?__ L
18. 1435 L = __?__ kL
19. 0.5 g = __?__ mg
20. 8346 g = __?__ kg
21. 1756 mg = __?__ g
22. 0.25 kL = __?__ L
23. 476 L = __?__ kL
24. 4100 mm = __?__ cm

Complete.

25. To convert from liters to milliliters, —————————.

26. To convert from liters to kiloliters, —————————.

27. Multiply when you are converting from meters to —————————.

28. Divide when you are converting from meters to —————————.

Solve.

29. A bolt of fabric is 2 m long. Charles cuts a 25 cm length from the bolt of fabric. How many centimeters of fabric are left on the bolt?

30. You pour 6 cans of juice containing 200 mL each and 3 cans of juice containing 750 mL each into a punchbowl. How many liters of juice are in the bowl?

Topic 4 *Warm-ups*

Standardized Testing Warm-Ups

1. Which distance is the shortest?

   A 100,000 mm   **B** 60,000 cm   **C** 700 m   **D** 0.56 km

2. How many kiloliters are equivalent to 345,000 milliliters?

   A 345   **B** 345,000,000   **C** 34.5   **D** 0.345

Homework Review Warm-Ups

3. Mario has 379 mL of water left in his 1-liter container. How much has he already had to drink?

Express each measurement in terms of meters.

4. 453 cm   5. 9562 mm   6. 0.345 km   7. 4 km

Topic 4   3   Using Formulas to Find Perimeter and Circumference

GOAL

Find the perimeters of polygons. Find the circumference of a circle.

Marianne is putting a border around the outside of her vegetable garden. To determine the total length of border she needs, Marianne can measure the perimeter of the garden.

Terms to Know

Example / Illustration

Terms to Know	Example / Illustration
Perimeter the distance around a polygon	3.5 yd 3.25 yd 3.25 yd · Garden · 1.75 yd 6.25 yd The perimeter of the garden is 3.25 + 3.5 + 3.25 + 1.75 + 6.25 = 18 yd.
Radius (*r*) the length of any segment that has one endpoint at the center of a circle and the other endpoint on the circle	$C \approx 31.4$ cm $r = 5$ cm $d = 10$ cm
Diameter (*d*) the distance across a circle through its center	The radius of the circle is 5 cm. The diameter of the circle is 10 cm.
Circumference (*C*) the distance around a circle	The circumference of the circle is about 31.4 cm.
Pi (π) the number that is the ratio of the circumference of any circle to its diameter	$\pi \approx 3.14159265\ldots$ For most purposes, you can use the approximation $\pi \approx 3.14$.
Regular a polygon with all sides the same length and all angle measures equal	(square)

UNDERSTANDING THE MAIN IDEAS

To find the perimeter of any polygon, find the sum of the lengths of its sides. If a polygon is regular, then you can multiply the measure of one side by the number of sides.

Example 1

Find the perimeter of each polygon.

a.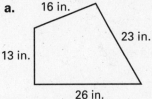

b.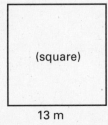

c.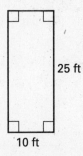

■ Solution ■

a. The quadrilateral is not a regular polygon. The sides are different lengths. Add the length of each side to find the perimeter.

$$13 + 16 + 23 + 26 = 78$$

The perimeter is 78 in.

b. A square is a regular quadrilateral. All sides are the same length. To find the perimeter, multiply the length of one side by the number of sides.

$$4 \cdot 13 = 52$$

The perimeter is 52 m.

c. The polygon is a quadrilateral with four right angles, so it is a rectangle. A rectangle is not a regular polygon, but its opposite sides have the same lengths. The perimeter is twice the length plus twice the width.

Perimeter of a rectangle = $2\ell + 2w$

For the rectangle shown,

$$P + 2\ell + 2w = (2 \cdot 25) + (2 \cdot 10) = 50 + 20 = 70$$

The perimeter of the rectangle is 70 ft.

Find the perimeter of each polygon. If only one side length is given, assume that the polygon is regular.

1.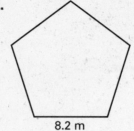

8.2 m

2.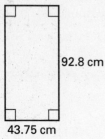

92.8 cm

43.75 cm

3.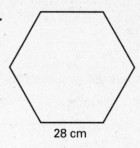

28 cm

4.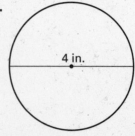

5 ft 5 ft

5.5 ft 5.5 ft

4.5 ft

5.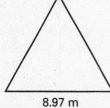

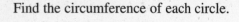

8.97 m

6.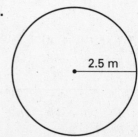

1.5 m 3.9 m

3.6 m

7. The Pentagon is one of the largest office buildings in the United States. It has the shape of a regular pentagon with a side length of about 281 m. Find its perimeter.

If you know a circle's diameter, there is a simple formula for its circumference.

Circumference of a circle = πd

Because the diameter of a circle is twice the radius, you can also write the formula in the following form.

Circumference of a circle = $2\pi r$

Example 2

Find the circumference of each circle.

a.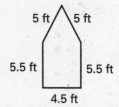

4 in.

b.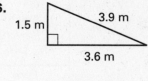

2.5 m

■ Solution ■

Step 1: **a.** The diameter is given.
Use the formula $C = \pi d$.

b. The radius is given.
Use the formula $C = 2\pi r$.

Step 2: **a.** Substitute numbers for the
letters in the formula.
Then solve. (Use 3.14 for π.)

$$C = \pi d$$
$$\approx 3.14 \cdot 4$$
$$= 12.56$$

The circumference of the
circle is about 12.56 in.

b. Substitute numbers for the
letters in the formula.
Then solve. (Use 3.14 for π.)

$$C = 2\pi r$$
$$\approx 2 \cdot 3.14 \cdot 2.5$$
$$= 15.7$$

The circumference of the
circle is about 15.7 m.

Find the circumference of each circle. Use 3.14 for π.

8.

15 in.

9.
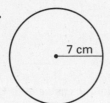
7 cm

10.
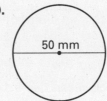
50 mm

11.
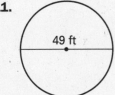
49 ft

12.
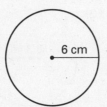
6 cm

13.
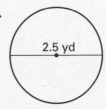
2.5 yd

Estimate whether the circumference is greater than or less than 100 ft.

14. $d = 35$ ft

15. $d = 25$ ft

16. $d = 30$ ft

· · · · · · · · · · · · · · · ·
Spiral Review

Complete.

17. 378 in. = __?__ yd

18. 214.5 lb = __?__ oz

19. 6 qt = __?__ fl oz

20. Write the fraction $\dfrac{4}{25}$ as a decimal and as a percent.

Topic 4   3   *Practice*

Find the perimeter of each polygon. If only one side length is given, assume that the polygon is regular.

1.
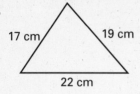
17 cm   19 cm
22 cm

2.
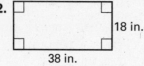
18 in.
38 in.

3.
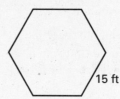
15 ft

4.
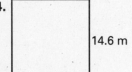
14.6 m

5.
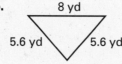
8 yd
5.6 yd   5.6 yd

6.
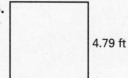
4.79 ft

7.
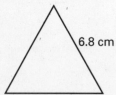
6.8 cm

8.
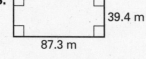
39.4 m
87.3 m

9.

9.23 ft

Find the circumference of each circle. Use 3.14 for π. Round answers to the nearest hundredth.

10.
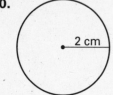
2 cm

11.
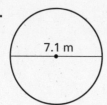
7.1 m

12.
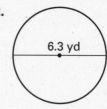
6.3 yd

13.
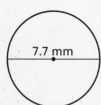
7.7 mm

14.
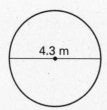
4.3 m

15.
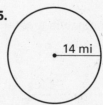
14 mi

Solve.

16. Suzanne is making a frame for an 8-inch by 10-inch photo. She has a yard of wood. Is that enough to make the frame? Explain.

17. The diameter of a quarter is 24 mm. What is the quarter's circumference?

Standardized Testing Warm-Ups

1. A regular *nonagon* has nine sides. If one of the sides has a length of 12 mm, what is the perimeter?

   A 42 mm   **B** 108π mm   **C** 98 mm   **D** 108 mm

2. A circle has a radius of 30 m. What is the best estimate for its circumference?

   A 90 m   **B** 94 m   **C** 180 m   **D** 188 m

Homework Review Warm-Ups

3. Which has a greater perimeter, a square with sides measuring 1.2 m or a rectangle measuring 2.2 m long and 1.1 m wide?

4. Each side of a regular octagon is 4.65 m long. What is its perimeter?

5. A flying disk has a diameter of 9 in. What is its circumference? Use 3.14 for π.

Topic 4

Using Formulas to Find Area

GOAL

Use formulas to find the areas of rectangles, squares, parallelograms, trapezoids, and triangles.

Joseph is redecorating his kitchen. He is laying new tiles on the rectangular floor, which measures 6 feet by 10 feet. The perimeter of the kitchen is $2 \cdot 6 + 2 \cdot 10 = 32$ feet. But to know how much tile to buy, Joseph instead needs to know the *area* of the floor.

Terms to Know

Example / Illustration

Square unit a measurement unit that is exactly one unit long and one unit wide; A square unit may be a square inch, a square centimeter, a square mile, and so on.	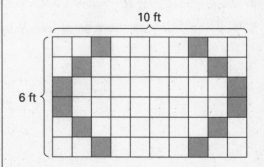 1 ft 1 ft Each tile measures 1 square foot.
Area the number of square units needed to cover a figure in a plane	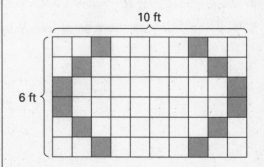 10 ft 6 ft It takes 60 tiles to cover the floor. The area is 60 square feet.

UNDERSTANDING THE MAIN IDEAS

There are formulas for finding the areas of many different polygons. For the kitchen floor, notice that the area is 60 square feet, and the length times the width is $6 \times 10 = 60$. You can also think of the length and width in terms of a *base* and a *height*.

Area of a rectangle: Area = base × height, or $A = b \cdot h$

Area of a parallelogram: Area = base × height, or $A = b \cdot h$

A square is just a rectangle with an equal base and height, or four sides of equal length (s).

Area of a square: Area = side × side, or $A = s \cdot s = s^2$

Remember that you always give an area in square units, or units^2.

Example 1

Find the area of each figure.

a.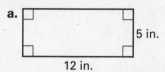
5 in.
12 in.

b.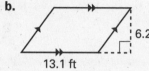
6.2 ft
13.1 ft

c.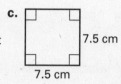
7.5 cm
7.5 cm

■ Solution ■

a. The figure is a rectangle with a base of 12 in. and a height of 5 in. Use these numbers in the formula. The answer will have units of square inches.

$$A = b \cdot h = 12 \cdot 5 = 60$$

The area of the rectangle is 60 in.^{2}.

b. The figure is a parallelogram with a base of 13.1 ft and a height of 6.2 ft. Use these numbers in the formula. The height is the shortest distance from the base to the top; it is not actually the length of a side of this parallelogram. The answer will have units of square feet.

$$A = b \cdot h = 13.1 \cdot 6.2 = 81.22$$

The area of the parallelogram is 81.22 ft^2.

c. The figure is a square with a side length of 7.5 cm. Use this number in the formula. The answer will have units of square centimeters.

$$A = s^2 = 7.5^2 = 7.5 \cdot 7.5 = 56.25$$

The area of the square is 56.25 cm^2.

Find the area of each rectangle, square, or parallelogram. Remember to express your answer in square units.

1.
2.1 cm
4.8 cm

2.
2.4 in.
3.8 in.

3.
26 in.

4.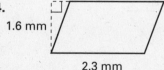
1.6 mm
2.3 mm

5.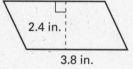
21 cm
42 cm

6.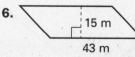
15 m
43 m

Imagine cutting a parallelogram in half between opposite corners. The result is two congruent triangles. The area of each triangle is just half the area of the parallelogram.

Area of a triangle: Area $= \frac{1}{2} \times$ base $\times$ height, or $A = \frac{1}{2} \cdot b \cdot h$

Example 2

Find the area of each triangle.

a.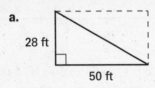

28 ft

50 ft

b.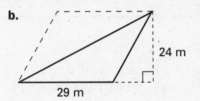

24 m

29 m

▪ Solution ▪

a. You can see that the area of the triangle is one half the area of the rectangle with the same base and height. The triangle has a base of 50 ft and a height of 28 ft. Use these numbers in the formula.

$$A = \frac{1}{2} \cdot b \cdot h = \frac{1}{2} \cdot 50 \cdot 28 = 700$$

The area of the triangle is 700 ft^2.

b. You can see that the area of the triangle is one half the area of the parallelogram with the same base and height. The triangle has a base of 29 m and a height of 24 m. Use these numbers in the formula.

$$A = \frac{1}{2} \cdot b \cdot h = \frac{1}{2} \cdot 29 \cdot 24 = 348$$

The area of the triangle is 348 m^2.

Calculate the area of each triangle. Use the formula $A = \frac{1}{2} \cdot b \cdot h$.

7.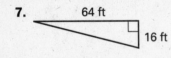
64 ft
16 ft

8.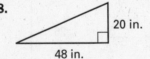
20 in.
48 in.

9.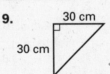
30 cm
30 cm

10.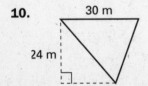
30 m
24 m

11.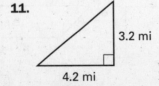
3.2 mi
4.2 mi

12.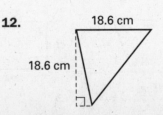
18.6 cm
18.6 cm

Recall that a *trapezoid* is a quadrilateral with exactly one pair of parallel sides. Both of the parallel sides of a trapezoid are called *bases*. You write them as b_1 and b_2. The area of a trapezoid is just the average length of the bases, $\frac{1}{2}(b_1 + b_2)$, times the height of the trapezoid.

Area of a trapezoid: Area $= \frac{1}{2} \times$ sum of base lengths $\times$ height, or $A = \frac{1}{2}(b_1 + b_2) \cdot h$

Example 3

Find the area of the polygon.

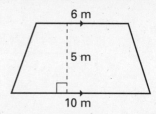

■ Solution ■

The figure is a trapezoid with a height of 5 m. Its two bases have lengths of 6 m and 10 m. Use these numbers in the formula. It doesn't really matter which base length you use for b_1 and which you use for b_2. The answer will have units of square meters.

$$A = \frac{1}{2}(b_1 + b_2) \cdot h = \frac{1}{2}(6 + 10) \cdot 5 = 8 \cdot 5 = 40$$

The area of the trapezoid is 40 m².

Find the area of each trapezoid. Use the formula $A = \frac{1}{2}(b_1 + b_2) \cdot h$.

13.

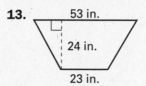

14.

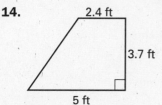

15.

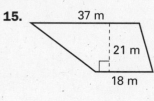

....
Spiral Review

16. Order the quantities 4.5 pt, 75 fl oz, 2 qt, and 9.5 c from greatest to least.

17. Evaluate $|x + 2| - 1$ when $x = -3$.

Find the area of each polygon.

1.
5 mm
24 mm

2.
4 ft
8 ft

3.
2.5 yd
2.5 yd

4.
24 m
12 m
10 m

5.
4.5 cm
6.8 cm

6.
18 m
12 m

7.
5.2 in.
4.8 in.

8.
1.4 m
2.1 m

9.
3.4 mi
3.4 mi

10.
10 m
10 m
20 m

11.
4.8 cm
6.2 cm

12.
6 cm
8.7 cm

13.
7.2 m
6.5 m

14.
9.8 mm
10 mm
12.6 mm

15.
4.9 m
3.8 m

Solve.

16. The height of a rectangular wall is 8 m less than its base. The base is 17 m. What is the height of the wall? What is its area?

17. A triangle and a rectangle have the same base and the same height. What do you know about their areas?

18. Of all rectangles with perimeter 36 cm, the one with the greatest area is a square. Find its dimensions and its area.

Standardized Testing Warm-Ups

1. What is the area of the parallelogram at the right?

   A 18 cm   **B** 36 cm^2

   C 72 cm^2 **D** 78 cm^2

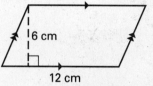

2. What is the area of the trapezoid at the right?

   A 45.5 in.^{2} **B** 77 in.^{2}

   C 91 in.^{2} **D** 105 in.^{2}

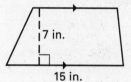

Homework Review Warm-Ups

Find the area of each polygon.

3.

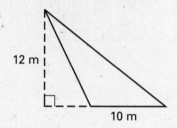

4.

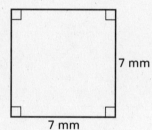

5.

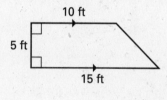

Topic 5 — LESSON 1 — *Bar Graphs and Line Graphs*

GOAL

Read and interpret bar graphs and line graphs. Construct a bar graph or a line graph for a data set.

Data may be displayed using a graph. The type of graph that is most effective depends on the nature of the data.

Terms to Know	*Example / Illustration*
Bar graph a type of graph in which the lengths of vertical or horizontal bars are used to compare data that fall into different categories	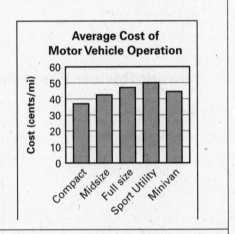
Line graph a type of graph in which data points connected by line segments show the change in the data over time	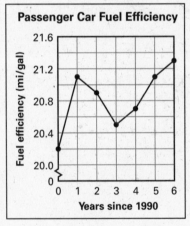

UNDERSTANDING THE MAIN IDEAS

One axis (most often the vertical axis) of a bar graph is marked with a numerical scale. The other axis is labeled with several categories. The data are represented by the lengths of the bars.

Example 1

Refer to the bar graph shown in the Terms to Know on the previous page. Use the graph to estimate the average operating cost of a minivan.

■ Solution ■

Locate the bar labeled "Minivan." The top of the bar is about halfway between 40 and 50 on the vertical scale. Therefore, the graph shows the cost of operating a minivan is about 45¢/mi.

Refer to the bar graph shown in the Terms to Know on the previous page.

1. Estimate the average cost of operating a midsize car.

2. Compare the operating costs of full-size cars and compact cars.

In many line graphs, one axis is a numerical scale. The other axis usually represents a period of time. A gap in a scale is indicated by a jagged line. The data in a line graph are represented by points connected by line segments.

Example 2

Use the line graph shown in the Terms to Know on the previous page. Estimate the average fuel efficiency for passenger cars in 1993.

■ Solution ■

Locate "1993" on the horizontal axis. Move straight up from "1993" until you reach the line. From this point, move directly left to the vertical axis and estimate the value on the scale: about 20.5.

The average fuel efficiency for passenger cars in 1993 was about 20.5 mi/gal.

Refer to the line graph shown in the Terms to Know on the previous page.

3. Estimate the average fuel efficiency for passenger cars in 1991.

4. Did the average fuel efficiency increase or decrease from 1993 to 1994 and by how much?

You can draw bar graphs to display data involving different categories.

Example 3

Draw a bar graph to display the data in the table.

Annual Snowfall at Colorado Ski Areas

Ski Area	Aspen Mt.	Bear Creek	Keystone	Telluride	Vail
Snowfall (in.)	300	330	230	300	335

■ Solution ■

Begin by writing the five ski areas along the horizontal axis. Then choose a numerical scale and label the vertical axis. Since the greatest data value is 335 inches, a scale from 0 inches to 350 inches with intervals of 50 inches is used here.

Now use the data in the table to draw a bar for each ski area. Finally, write a title for the graph.

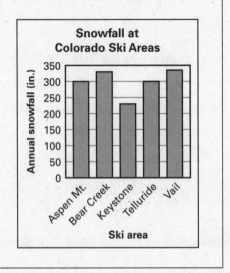

5. Only seven states are the birthplace of more than one United States president. The table below shows the number of presidents born in each of those seven states. Draw a bar graph to display the data.

State	Number of presidents
Massachusetts	4
North Carolina	2
New York	4
Ohio	7
Texas	2
Vermont	2
Virginia	8

You can draw a line graph to display data involving a single category for which the value changes over time.

Example 4

Draw a line graph to display the data in the table below.

U.S. Patent Applications

Year	1991	1992	1993	1994	1995
Number of applications (thousands)	178	185	188	201	236

■ Solution ■

Begin by writing the years along the horizontal axis. Then choose a numerical scale for the vertical axis. Since all of the data values are between 178 and 236, show a break in the scale after 0 and then show values from 170 to 270 using intervals of 20.

Now use the data in the table to plot a point for each year, and connect the five points from left to right with line segments. Finally, write a title for the graph.

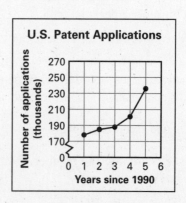

6. The table at the right shows the number of personal computers in use in the United States each year from 1991 to 1995. Draw a line graph to display the data.

7. Explain how the data in Exercise 6 can also be displayed in a bar graph.

Year	*Number of personal computers in the U.S. (nearest million)*
1991	59
1992	65
1993	73
1994	82
1995	92

· · · · · · · · · · · · · · · · · · · ·
Spiral Review

Multiply.

8. $(-5)(-11)$

9. $6(-7)$

10. $(-11)(11)$

Topic 5 **Practice**

For Exercises 1–6, use the bar graph at the right.

For Exercises 1–3, estimate the number of Calories used per minute for each activity.

1. walking at a rate of 3 mi/h

2. tennis   3. jogging

4. For which of the listed activities would a 120 lb person use more than 5 Calories per minute?

5. Which activity produces a Calorie use about twice that of volleyball?

6. About how many Calories would a 120 lb person use if he or she rode a bicycle at a rate of 5 mi/h for half an hour?

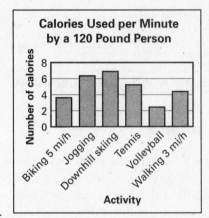

For Exercises 7–11, use the line graph.

For Exercises 7 and 8, estimate the number of computers per 1000 people each year.

7. 1992   8. 1994

9. In what year were there about 35 computers per 1000 people worldwide?

10. In what year did the number of computers per 1000 people pass 40 for the first time?

11. By approximately how much did the number of computers per 1000 people increase from 1991 to 1995?

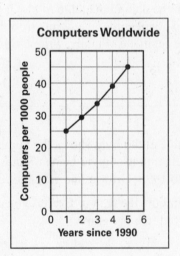

12. Draw a bar graph to display the data in the table.

Sites of the Olympic Games by Continent (1896–2002)

Continent	Asia	Australia	Europe	North America
Games held	4	2	27	11

13. Draw a line graph to display the data in the table.

Consumer Credit in the United States

Year	1990	1991	1992	1993	1994	1995
Amount owed (billions of dollars)	752	745	757	807	925	1132

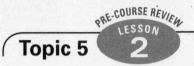

Topic 5   *Warm-ups*

Standardized Testing Warm-Ups

1. A nurse records a patient's temperature every hour for 24 h. Which type of graph would best display the data?

   A bar graph

   B circle graph

   C line graph

   D pictograph

Homework Review Warm-Ups

Use the bar graph at the right.

2. About how many mammal species are endangered?

3. Estimate the difference between the number of endangered bird species and the number of endangered reptile species.

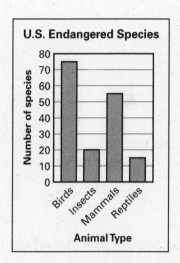

Use the line graph at the right.

4. Estimate the population of the United States in 1970.

5. By about how much did the population of the United States increase between 1950 and 1990?

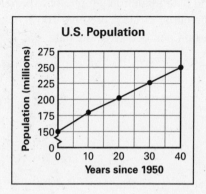

Topic 5   **2**   *Circle Graphs*

GOAL

Read and interpret
circle graphs.

A circle graph is a very useful display when you want to show how each
part of something relates to the whole thing, as well as to show how the
parts relate to each other.

Terms to Know	*Example / Illustration*
Circle Graph a type of graph in which the data are represented as parts of a whole	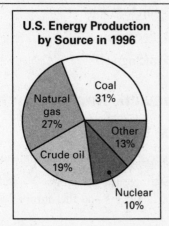 **U.S. Energy Production by Source in 1996**
Sector a wedge-shaped part of a circle; in a circle graph, a region that represents part of the data being displayed	The circle graph above contains five sectors.

UNDERSTANDING THE MAIN IDEAS

The data in a circle graph are expressed as parts (or percentages) of a whole.
The entire circle represents 100% of the data and each sector represents a
certain percentage of the data.

Example 1

Use the circle graph shown in the Terms to Know on the previous page. If the total amount of energy produced in the United States in 1996 was about 70 quadrillion British thermal units (Btu's), about how much energy was produced from natural gas?

■ Solution ■

The circle graph shows that the amount of energy produced in the United States in 1996 from natural gas was 27% of the total energy produced.

$$27\% \text{ of } 70 = 0.27 \times 70 = 18.9$$

About 18.9 quadrillion Btu's were produced in the United States from natural gas.

Use the circle graph in the Terms to Know on the previous page. Suppose the total United States energy production in 1996 was about 70 quadrillion Btu's.

1. About how much energy was produced in the United States from crude oil?

2. About how much energy was produced from fossil fuels, that is, coal, natural gas, and crude oil?

Circle graphs can also be drawn showing the actual data values in the sectors, rather than percents.

Example 2

Use the circle graph at the right. The graph shows the amount of energy of various types (in quadrillion Btu's) that was consumed in the United States in 1996.

a. What was the total amount of energy consumed in the United States in 1996?

b. What percent of the energy consumed in 1996 was accounted for by natural gas?

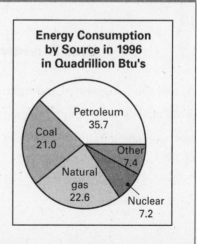

Energy Consumption by Source in 1996 in Quadrillion Btu's

Petroleum 35.7
Coal 21.0
Other 7.4
Natural gas 22.6
Nuclear 7.2

▪ Solution ▪

a. Add the amounts shown in the five sectors.

$$35.7 + 21.0 + 22.6 + 7.2 + 7.4 = 93.9$$

The total amount of energy consumed in the United States in 1996 was 93.9 quadrillion Btu's.

b. The total amount consumed was 93.9 quadrillion Btu's, while the total natural gas consumption was 22.6 quadrillion Btu's.

$$22.6 \text{ quadrillion out of } 93.9 \text{ quadrillion} = \frac{22.6 \text{ quadrillion}}{93.9 \text{ quadrillion}} = \frac{22.6}{93.9}$$

Divide to find decimal: $22.6 \div 93.9 \approx 0.241$

Then write as a percent: $0 \,.\, 2 \,\, 4 \,\, 1 \longrightarrow 24.1\%$

So, natural gas accounted for about 24.1% of the energy consumed in the United States in 1996.

Use the circle graph in Example 2 on the previous page. Find the percent of the total energy consumption in the United States in 1996 for each energy source. Round your answers to the nearest tenth of a percent.

3. nuclear

4. other

5. all fossil fuels (petroleum, natural gas, coal)

· · · · · · · · · · · · · · · · · · · ·

Spiral Review

In Exercises 6–8, use the bar graph at the right.

6. How many gold medals were won by Germany?

7. Which country won about half as many gold medals as the United States?

8. How many countries won more than 20 gold medals?

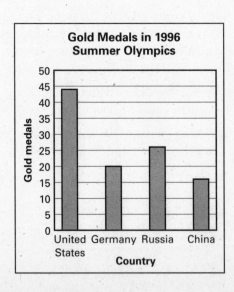

Gold Medals in 1996 Summer Olympics

Private health insurance falls into the four categories shown in the circle graph at the right. In Exercises 1–4, estimate the number of persons with each type of insurance among a group of 100,000 privately insured people.

1. private care

2. Health Maintenance Organization (HMO)

3. managed care—choice of doctor

4. managed care—specified doctor

5. In a group of 65,000 privately insured people, about 15,000 have managed care in which they may choose their own doctors. How does the percent of people with this type of insurance compare to the percent shown in the circle graph above?

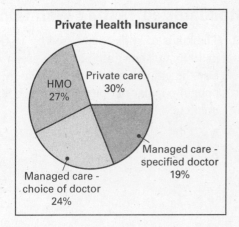

Private Health Insurance

HMO 27%
Private care 30%
Managed care - specified doctor 19%
Managed care - choice of doctor 24%

The number of motor vehicles registered in the United States in 1995 was about 206,000,000. In Exercises 6–8, use the circle graph at the right to estimate the number of each type of vehicle.

6. automobiles

7. trucks and buses (combined)

8. motorcycles

9. If a circle graph was drawn just for the number of motor vehicles registered in the state of Michigan, the percentages would be the same as those in the circle graph above. In 1995, there were about 8,137,000 registered motor vehicles in Michigan. Estimate the total number of automobiles registered in Michigan in 1995 to the nearest thousand.

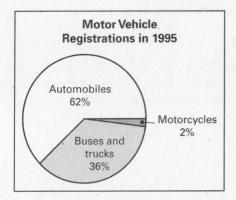

Motor Vehicle Registrations in 1995

Automobiles 62%
Motorcycles 2%
Buses and trucks 36%

Use the circle graph at the right.

10. Find the number of ski runs at Alpine Cliffs.

11. What percent of the ski runs at Alpine Cliffs is in each category?

   a. expert

   b. intermediate

   c. beginner

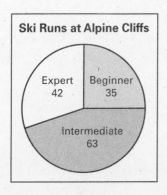

Ski Runs at Alpine Cliffs

Expert 42
Beginner 35
Intermediate 63

Topic 5 Warm-ups

Standardized Testing Warm-Ups

1. The kinds of books purchased by 16,000 adults in 1996 are shown in the circle graph at the right. Use the graph to find how many more people bought popular fiction books than bought general nonfiction books.

   A 6720   **B** 8160

   C 3360   **D** 7840

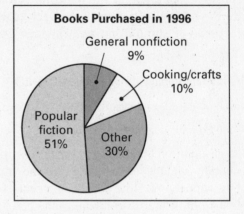

Books Purchased in 1996

General nonfiction 9%
Cooking/crafts 10%
Popular fiction 51%
Other 30%

Homework Review Warm-Ups

Use the circle graph at the right.

2. What percent of the forest land had forest on it?

3. There were 191 million acres of National Forest land in 1995. Estimate the number of acres on which timber harvest was permitted.

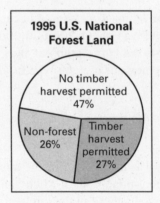

1995 U.S. National Forest Land

No timber harvest permitted 47%
Non-forest 26%
Timber harvest permitted 27%

Topic 5 *Interpreting Graphs*

GOAL

Recognize how the visual impression given by a graph can be misleading.

In Lesson 2.2, you learned how to display data using bar graphs and line graphs. Depending on how they are drawn, two bar graphs or two line graphs that display exactly the same data can create two different visual impressions. Some graphs could even be misleading to the person viewing them.

UNDERSTANDING THE MAIN IDEAS

The scale of a graph can be chosen in such a way as to make a desired impression on the viewer.

Example 1

Both of these bar graphs display the same data.

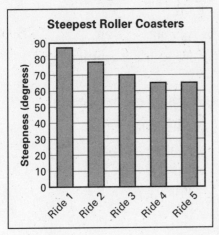

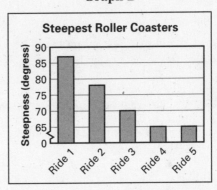

a. Explain why the visual impression given by Graph B could be misleading.

b. Explain why an advertisement about Ride 1 roller coaster would more likely include Graph B than Graph A.

Solution

a. Because of the scale on Graph B, the relative lengths of the bars are misleading. For example, the bar for the steepest roller coaster in the group, Ride 1, is five times as long as those for the two least steep roller coasters. A viewer who looks at the bar lengths without considering the numerical values of the bars might conclude that Ride 1 is five times as steep as Ride 4 and Ride 5, but it is actually much less than twice as steep.

b. The visual appearance of Graph B overemphasizes the steepness of Ride 1, making it seem more thrilling.

Use the graphs in Example 1.

1. Describe the visual impression that Graph B gives of the steepness of Ride 3 and Ride 5 roller coasters.

2. Explain why the owners of Ride 3 roller coaster might use Graph A in an advertisement.

The visual impression given by a line graph could be misleading if the choice of graph scale makes the change in the data over time appear to be more dramatic than it actually is.

Example 2

Both line graphs below show the median income for a family of four in the United States from 1988 to 1994.

Graph A

Graph B

a. Compare the visual impressions made by the two graphs.

b. Which graph would you use to support an argument that the economic situation in the United States improved greatly during the period from 1988 to 1994?

▪ Solution ▪

a. Graph C makes it appear that the median income for a family of four has improved slowly but steadily from 1988 to 1994. Graph D makes it appear that the median income for a family of four increased sharply over that same period of time.

b. Graph D; this graph makes the increases from one year to the next seem more dramatic than they appear in Graph C.

Use the graphs in Example 2.

3. Under what circumstances do you think a candidate running for reelection would use Graph D in his or her campaign literature?

4. Look at the labels on the horizontal axis of Graph C. What effect do you think spacing the labels farther apart would have on the visual impression given by the line graph?

Spiral Review

Gregory tries to balance his caloric intake from fat, protein, and carbohydrates as shown in the graph. His total intake is 2000 Calories per day.

5. How many Calories does he get from protein?

6. How many Calories does he get from carbohydrates?

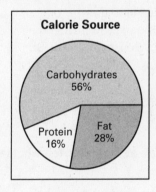

Calorie Source

Carbohydrates 56%

Protein 16%

Fat 28%

Topic 5 3 *Practice*

Use the bar graph at the right for Exercises 1–4.

1. What visual impression do the lengths of the bars give about the relationship between gourmet coffee prices at Alonzo's and at Cafe Break?

2. What visual impression do the lengths of the bars give about the relationship among gourmet coffee prices at all four stores?

3. Which store do you think is most likely to use this graph in its advertising? Which is least likely to use it?

4. Explain how the graph could be redrawn to give a more favorable visual impression of coffee prices at Cafe Break.

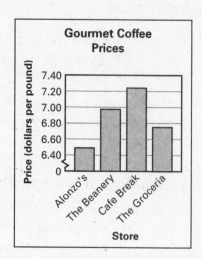

Use the line graphs at the right for Exercises 5 and 6.

5. Compare the visual impressions given by the two graphs.

6. A developer is trying to convince investors to increase the size of a planned supermarket. Which graph should the developer use? Explain.

Graph A

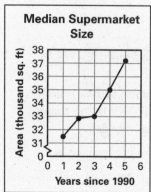

Graph B

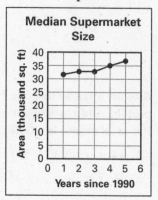

Use the bar graph at the right. Tell whether each statement is *true* or *false*.

7. The game *Star Traveler* costs almost twice as much at Comp-Town as it does at Romulus.

8. The game costs only half as much at CD Land as it does at Game Shack.

9. The difference between the highest and lowest prices for the game is $7.

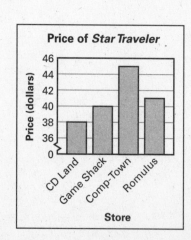

Part 2   Key Standards

Algebra 1

KEY STANDARD

2.0

Students understand and use such operations as taking the opposite, finding the reciprocal, taking a root, and raising to a fractional power. They understand and use the rules of exponents.

TEXTBOOK REFERENCES
Developing Concepts 8.1
Lesson 8.1

KEY WORDS

- power
- base
- exponent
- product of powers property
- power of a power property
- power of a product property

Multiplication Properties of Exponents

MULTIPLICATION PROPERTIES OF EXPONENTS

PRODUCT OF POWERS PROPERTY

To multiply powers with the same base, add the exponents.

For any number a and any integers m and n, $a^m \cdot a^n = a^{m+n}$.

POWER OF A POWER PROPERTY

To find a power of a power, multiply the exponents.

For any number a and any integers m and n, $(a^m)^n = a^{mn}$.

POWER OF A PRODUCT PROPERTY

To find a power of a product, find the power of each factor and multiply.

For any numbers a and b and any integer m, $(ab)^m = a^m b^m$.

My Notes

Example 1

Write the expression as a single power of the base.

a. $7^3 \cdot 7^2$ **b.** $4 \cdot 4^4$ **c.** $b^3 \cdot b^4 \cdot b^6$ **d.** $(-3)(-3)^5$

▶ **Solution**

a. $7^3 \cdot 7^2 = 7^{3+2} = 7^5$ **b.** $4 \cdot 4^4 = 4^1 \cdot 4^4 = 4^{1+4} = 4^5$

c. $b^3 \cdot b^4 \cdot b^6 = b^{3+4+6}$ **d.** $(-3)(-3)^5 = (-3)^1 \cdot (-3)^5$
$= b^{13}$ $= (-3)^{1+5} = (-3)^6$

Example 2

Write the expression as a single power of the base.

a. $(x^3)^6$ **b.** $c^5(c^3)^2$

▶ **Solution**

a. $(x^3)^6 = x^{3 \cdot 6} = x^{18}$

b. $c^5(c^3)^2 = c^5 \cdot c^{3 \cdot 2} = c^5 \cdot c^6 = c^{5+6} = c^{11}$

Example 3

Simplify the expression.

a. $(4 \cdot 3)^4$ **b.** $(-2a)^2$

▶ **Solution**

a. $(4 \cdot 3)^4 = 4^4 \cdot 3^4$ Raise each factor to a power.

$= 256 \cdot 81$ Evaluate each power.

$= 20{,}736$ Multiply.

b. $(-2a)^2 = (-2 \cdot a)^2$ Identify factors.

$= (-2)^2 \cdot a^2$ Raise each factor to a power.

$= 4a^2$ Evaluate power.

SUMMARIZING KEY IDEAS
You can use the multiplication properties of exponents to simplify exponential expressions.

Checkpoint ✓ **Multiplication Properties of Exponents**

Write the expression as a single power of the base.

1. $4^3 \cdot 4^5$ **2.** $x^4 \cdot x^3 \cdot x^2$ **3.** $(a^4)^2$ **4.** $[(-2)^6]^4$

Simplify the expression.

5. $(-3a)^2$ **6.** $(2xy)^3$ **7.** $-(3a^5)^2$

Exercises **Multiplication Properties of Exponents**

Write the expression as a single power of the base.

1. $5^2 \cdot 5^3$ **2.** $6^3 \cdot 6^4$ **3.** $10 \cdot 10^8$ **4.** $b^3 \cdot b^{10} \cdot b^2$

5. $(5^4)^5$ **6.** $(x^3)^{15}$ **7.** $[(-3)^2]^4$ **8.** $-(w^3)^9$

TEXTBOOK LINK

On pages S2–S3, you used the multiplication properties of exponents to evaluate expressions. In your textbook, Lesson 8.1, you will use the product of powers property, the power of a power property, and the power of a product property to simplify expressions.

Simplify the expression.

9. $(2x)^3$ **10.** $(2x \cdot 3)^3$ **11.** $(3 \cdot 7)^4$ **12.** $(16 \cdot 2)^2$

13. $(-5a)^2$ **14.** $(5x^3)^6$ **15.** $(4a)^2 \cdot a$ **16.** $6^2 \cdot (6x^3)^2$

17. $2x^3 \cdot (3x)^2$ **18.** $(-ab)(a^2b)^2$ **19.** $(-rs)(rs^3)^2$ **20.** $(-3cd)^3(-d^2)$

Simplify the expression. Evaluate when $a = 1$ and $b = 2$.

21. $b^3 \cdot b^4$ **22.** $(a^2b)^5$ **23.** $(b^2 \cdot b^3) \cdot (b^2)^4$

TEXTBOOK REFERENCES
Lesson 8.2

KEY WORDS
• zero exponent
• negative exponent

Zero and Negative Exponents

> **ZERO AND NEGATIVE EXPONENTS**
>
> Let a be a nonzero number and let n be an integer.
> • A nonzero number to the zero power is 1. $a^0 = 1$, $a \neq 0$
> • The expression a^{-n} is the reciprocal of a^n. $a^{-n} = \dfrac{1}{a^n}$, $a \neq 0$

Example 1

Evaluate the expression.

a. 3^{-2} **b.** $(-4)^0$ **c.** m^{-3} **d.** $\dfrac{1}{5^{-1}}$ **e.** 0^{-2}

▶ **Solution**

a. $3^{-2} = \dfrac{1}{3^2} = \dfrac{1}{9}$ 3^{-2} is the reciprocal of 3^2.

b. $(-4)^0 = 1$ a^0 is equal to 1, $a \neq 0$.

c. $m^{-3} = \dfrac{1}{m^3}$ m^{-3} is the reciprocal of m^3, $m \neq 0$.

d. $\dfrac{1}{5^{-1}} = 5$ 5^1 is the reciprocal of 5^{-1}.

e. $0^{-2} = \dfrac{1}{0^2}$ (Undefined) a^{-n} is defined only for a *nonzero* number a.

▶**STUDY TIP**
Informally, you can think of rewriting expressions with positive exponents as "moving factors" from the denominator to the numerator and/or vice versa.

Example 2

Rewrite the expression with positive exponents.
a. $6(5^{-7})$ **b.** $4x^{-2}y^{-4}$

▶ **Solution**

a. $6(5^{-7}) = 6\left(\dfrac{1}{5^7}\right)$ **b.** $4x^{-2}y^{-4} = 4 \cdot \dfrac{1}{x^2} \cdot \dfrac{1}{y^4}$

       $= \dfrac{6}{5^7}$ $= \dfrac{4}{x^2 y^4}$

My Notes

Checkpoint ✓ *Zero and Negative Exponents*

Evaluate the expression.

1. 5^{-2} **2.** $(-14)^0$ **3.** $\dfrac{1}{3^{-3}}$ **4.** $\dfrac{1}{(-4)^{-3}}$

Rewrite the expression with positive exponents.

5. $7\left(4^{-5}\right)$ **6.** $6x^{-3}y^{-4}$ **7.** $\dfrac{9}{x^{-2}}$

Example 3

Evaluate the expression.

a. $4^{-2} \cdot 4^2$ **b.** $\left(3^{-3}\right)^{-2}$ **c.** 4^{-4}

▶ *Solution*

a. $4^{-2} \cdot 4^2 = 4^{-2+2} = 4^0 = 1$

b. $\left(3^{-3}\right)^{-2} = 3^{-3 \cdot (-2)} = 3^6 = 729$

c. $4^{-4} = \dfrac{1}{4^4} = \dfrac{1}{256}$

Example 4

Simplify the expression. Rewrite with positive exponents.

a. $(6a)^{-2}$ **b.** $\dfrac{1}{c^{-4n}}$

▶ *Solution*

a. $(6a)^{-2} = 6^{-2} \cdot a^{-2}$ Use the power of a product property.

$= \dfrac{1}{6^2} \cdot \dfrac{1}{a^2}$ Write reciprocals of 6^2 and a^2.

$= \dfrac{1}{36a^2}$ Multiply fractions.

b. $\dfrac{1}{c^{-4n}} = \left(c^{-4n}\right)^{-1}$ Use definition of negative exponents.

$= c^{(-4n) \cdot (-1)}$ Use power of a power property.

$= c^{4n}$ Multiply exponents.

SUMMARIZING KEY IDEAS

You can use the multiplication properties of exponents when working with zero and negative exponents.

Checkpoint ✓ **Zero and Negative Exponents**

Evaluate the expression.

8. $5^{-2} \cdot 5^2$ **9.** $(4^{-3})^{-2}$ **10.** 3^{-5} **11.** $(2 \cdot 6)^{-2}$

Rewrite the expression with positive exponents.

12. $(7b)^{-2}$ **13.** $\dfrac{1}{a^{-5n}}$ **14.** $\dfrac{12b^{-3}}{a^{-4}}$

Exercises **Zero and Negative Exponents**

Rewrite the expression using positive exponents.

1. x^{-7} **2.** $5x^{-4}$ **3.** $3x^{-2}$

4. $\dfrac{1}{2x^{-3}}$ **5.** $x^{-2}y^3$ **6.** x^6y^{-7}

7. $3x^{-3}y^{-8}$ **8.** $6x^{-2}y^{-4}$ **9.** $\dfrac{1}{7x^{-4}y^{-1}}$

Evaluate the expression.

10. 3^{-2} **11.** 2^{-4} **12.** -4^0

13. $6^3 \cdot 6^{-1}$ **14.** $8^4 \cdot 8^{-4}$ **15.** $(5^{-3})^2$

16. $-6 \cdot (-6)^{-1}$ **17.** $5 \cdot 5^{-1}$ **18.** $2^0 \cdot 3^{-3}$

Rewrite the expression with positive exponents.

19. $(-3)^0x$ **20.** $(5y)^{-2}$ **21.** $(-2x)^{-3}$

22. $[(-4)^0a]^{-2}$ **23.** $(-3x)^{-1} \cdot 2y$ **24.** $\dfrac{4}{b^{-2}}$

25. $(4xy)^{-2}$ **26.** $(2a^{-3})^3$ **27.** $\dfrac{1}{(4x)^{-3}}$

28. $\dfrac{9k^{-2}}{j^{-1}}$ **29.** $5^0x^{-3}y^{-2}z^{-5}$ **30.** $\dfrac{3}{q^{-3p}}$

TEXTBOOK LINK

📖 On pages S4–S6, you rewrote and evaluated expressions using zero and negative exponents. In your textbook, Lesson 8.2, you will evaluate and simplify powers that have zero and negative exponents and in Lesson 8.3 you will graph exponential functions. In Lesson 12.4, you will evaluate expressions involving rational exponents.

TEXTBOOK REFERENCES

Lesson 8.4

KEY WORDS

• quotient of powers property
• power of a quotient property

Division Properties of Exponents

DIVISION PROPERTIES OF EXPONENTS

Let a and b be numbers and let m and n be integers.

QUOTIENT OF POWERS PROPERTY

To divide powers with the same base, subtract the exponents.

$$\frac{a^m}{a^n} = a^{m-n}, a \neq 0$$

POWER OF A QUOTIENT PROPERTY

To find a power of a quotient, find the power of the numerator and the power of the denominator and divide.

$$\left(\frac{a}{b}\right)^m = \frac{a^m}{b^n}, b \neq 0$$

Example 1

Evaluate the expression.

a. $\dfrac{8^5}{8^4}$

b. $\dfrac{(-7)^2}{(-7)^2}$

c. $\dfrac{1}{x^4} \cdot x^2$

d. $\left(\dfrac{3}{7}\right)^2$

e. $\left(-\dfrac{4}{y}\right)^3$

f. $\left(\dfrac{8}{5}\right)^{-3}$

▶ **Solution**

a. $\dfrac{8^5}{8^4} = 8^{5-4} = 8^1 = 8$

b. $\dfrac{(-7)^2}{(-7)^2} = (-7)^{2-2} = (-7)^0 = 1$

c. $\dfrac{1}{x^4} \cdot x^2 = \dfrac{x^2}{x^4} = x^{2-4} = x^{-2} = \dfrac{1}{x^2}$

d. $\left(\dfrac{3}{7}\right)^2 = \dfrac{3^2}{7^2} = \dfrac{9}{49}$

e. $\left(-\dfrac{4}{y}\right)^3 = \left(\dfrac{-4}{y}\right)^3 = \dfrac{(-4)^3}{y^3} = \dfrac{-64}{y^3}$

f. $\left(\dfrac{8}{5}\right)^{-3} = \dfrac{8^{-3}}{5^{-3}} = \dfrac{5^3}{8^3} = \dfrac{125}{512}$

My Notes

Checkpoint ✓ **Division Properties of Exponents**

Evaluate the expression.

1. $\dfrac{6^3}{6^4}$ **2.** $\dfrac{(-6)^2}{(-6)^2}$ **3.** $\dfrac{6^4 \cdot 6^2}{6^7}$

4. $\left(\dfrac{2}{5}\right)^2$ **5.** $\left(-\dfrac{5}{y}\right)^3$ **6.** $\left(\dfrac{6}{4}\right)^{-3}$

Example 2

Simplify the expression. Use only positive exponents.

a. $\dfrac{3x^2y}{6x} \cdot \dfrac{6xy^2}{y^4}$ **b.** $\left(\dfrac{4x}{y^2}\right)^4$

▶ **Solution**

a. $\dfrac{3x^2y}{6x} \cdot \dfrac{6xy^2}{y^4} = \dfrac{(3x^2y)(6xy^2)}{(6x)(y^4)}$ Multiply fractions.

$\qquad\qquad\qquad = \dfrac{18x^3y^3}{6xy^4}$ Use product of powers property.

$\qquad\qquad\qquad = 3x^2y^{-1}$ Use quotient of powers property.

$\qquad\qquad\qquad = \dfrac{3x^2}{y}$ Use definition of negative exponents.

b. $\left(\dfrac{4x}{y^2}\right)^4 = \dfrac{(4x)^4}{(y^2)^4}$ Use power of a quotient property.

$\qquad\qquad\quad = \dfrac{4^4 \cdot x^4}{y^{2\,\cdot\,4}}$ Use power of a product property.
Use power of a power property.

$\qquad\qquad\quad = \dfrac{256x^4}{y^8}$ Evaluate power.
Multiply exponents.

SUMMARIZING KEY IDEAS

To divide powers that have the same base, you subtract exponents. When you have reduced an expression as far as possible and when all of your exponents are positive, the expression is in simplest form.

Checkpoint ✓ **Simplify Expressions**

Simplify the expression. Use only positive exponents.

7. $\dfrac{4x^2y}{5x} \cdot \dfrac{5xy^2}{y^4}$ **8.** $\left(\dfrac{3x}{y^4}\right)^3$ **9.** $\dfrac{a}{b^{-2}} \cdot \left(\dfrac{a^3}{b}\right)^{-2}$

TEXTBOOK LINK

On pages S7–S9, you used the division properties of exponents to evaluate and simplify expressions. In your textbook, Lesson 8.4, you will use the quotient of powers property and the power of a quotient property to divide with exponents.

Exercises Division Properties of Exponents

Simplify the expression. Use only positive exponents.

1. $\dfrac{6^6}{6^4}$
2. $\dfrac{8^3}{8^4}$
3. $\dfrac{(-4)^5}{(4)^5}$
4. $\dfrac{(-3)^9}{(-3)^9}$

5. $\dfrac{2^2}{2^{-3}}$
6. $\dfrac{8^3 \cdot 8^2}{8^5}$
7. $\dfrac{7^4 \cdot 7}{7^7}$
8. $\left(\dfrac{3}{4}\right)^2$

9. $\left(\dfrac{5}{3}\right)^3$
10. $\left(-\dfrac{2}{3}\right)^3$
11. $\left(-\dfrac{4}{5}\right)^2$
12. $\left(\dfrac{9}{6}\right)^{-1}$

13. $\left(\dfrac{2}{x}\right)^4$
14. $\dfrac{x^4}{x^5}$
15. $x^3 \cdot \dfrac{1}{x^2}$
16. $x^7 \cdot \dfrac{1}{x^9}$

17. $\dfrac{3x^2y^2}{3xy} \cdot \dfrac{6xy^3}{3y}$
18. $\dfrac{4xy^3}{2y} \cdot \dfrac{5xy^{-4}}{x^2}$
19. $\dfrac{16x^3y}{-4xy^3} \cdot \dfrac{-2xy}{-x}$

20. $\left(\dfrac{3^0b}{5}\right)^2$
21. $\left(\dfrac{5m^3n}{m^5}\right)^3$
22. $\dfrac{(3a^3)^2}{10a^{-1}}$

My Review Questions

Topic Review Rules of Exponents

These exercises will help you check that you can use the multiplication and division properties of positive, zero, and negative exponents to evaluate powers and simplify expressions. If you have any questions about the rules of exponents, be sure to get them answered before going on to the next section.

Evaluate the expression.

1. $6^2 \cdot 6^0$
2. $4^3 \cdot 4^{-4}$
3. $\left(\dfrac{5}{2}\right)^{-2}$
4. $(3^2)^3$

Simplify the expression. Use only positive exponents.

5. $q^3 \cdot q^2$
6. $x^2 \cdot x^2 \cdot x^3$
7. $(n^5)^2$
8. $-(t^3)^2$

9. $-3s^2 \cdot 8s^3$
10. $5x^{-5}$
11. $(-2b)^{-2}$
12. 4^0c^{-3}

13. $\dfrac{4x^2y}{2xy^3}$
14. $\dfrac{p^{-3}}{p^2}$
15. $\dfrac{4q^4}{3q^{-3}}$
16. $\left(\dfrac{2xy^4}{x^2}\right)^3$

17. $\dfrac{4w^2z^4}{5y^3z^2} \cdot \dfrac{10z^2}{2wy}$
18. $\dfrac{-9x^5y^7}{x^2y^3} \cdot \dfrac{(2xy)^2}{-6x^2y^2}$
19. $\dfrac{3a^{-2}b^{-1}}{a^{-3}} \cdot \dfrac{5a^{-4}}{b^3}$

Students simplify expressions before solving linear equations and inequalities in one variable, such as $3(2x - 5) + 4(x - 2) = 12$.

TEXTBOOK REFERENCES
Lesson 3.3

KEY WORDS
- multi-step equation
- inverse operations
- like terms
- distributive property

Linear Equations in One Variable

SIMPLIFYING BEFORE SOLVING AN EQUATION

An equation may have like terms on one or both sides of the equal sign. To solve the equation, you want to isolate the variable.

Example 1

Solve $3n + 9 + 4n = 2$.

▶ **Solution**

$3n + 9 + 4n = 2$	Write original equation.
$7n + 9 = 2$	Combine like terms $3n$ and $4n$.
$7n = -7$	Subtract 9 from each side.
$n = -1$	Divide each side by 7.

▶ The solution is -1.

Check ✓   Check by substituting -1 for each n in the *original* equation.

$3n + 9 + 4n = 2$	Write original equation.
$3(-1) + 9 + 4(-1) \stackrel{?}{=} 2$	Substitute -1 for each n.
$2 = 2$ ✓	Solution is correct.

> **STUDY TIP**
> The first step in solving equations is often to combine like terms.

Example 2

Solve $2(x + 7) - 4x = 8$.

▶ **Solution**

$2(x + 7) - 4x = 8$	Write original equation.
$2x + 14 - 4x = 8$	Use distributive property.
$-2x + 14 = 8$	Combine like terms $2x$ and $-4x$.
$-2x = -6$	Subtract 14 from each side.
$x = 3$	Divide each side by -2.

▶ The solution is 3. Check this in the original equation.

> **STUDY TIP**
> If an equation involves parentheses, you may need to use the distributive property before combining like terms.

My Notes

Example 3

Solve $4n - 7(n - 9) = 42$.

▶ **Solution**

$4n - 7(n - 9) = 42$	Write original equation.
$4n - 7n - 7(-9) = 42$	Use distributive property.
$4n - 7n + 63 = 42$	Multiply -7 and -9.
$-3n + 63 = 42$	Combine like terms.
$-3n = -21$	Subtract 63 from each side.
$n = 7$	Divide each side by -3.

▶ The solution is 7. Check this in the original equation.

Example 4

Solve $66 = -\dfrac{6}{5}(x + 3)$.

▶ **Solution**

$66 = -\dfrac{6}{5}(x + 3)$	Write original equation.
$66 = \dfrac{-6}{5}(x + 3)$	Move the negative sign into the numerator.
$\left(\dfrac{5}{-6}\right)(66) = \left(\dfrac{5}{-6}\right)\left(\dfrac{-6}{5}\right)(x + 3)$	Multiply each side by $\dfrac{5}{-6}$.
$-55 = x + 3$	Simplify.
$-58 = x$	Subtract 3 from each side.

▶ The solution is -58. Check this in the original equation.

▶**STUDY TIP**
In Example 4, you can clear the equation of fractions by multiplying by the reciprocal of $\dfrac{-6}{5}$.

Checkpoint ✓ **Simplify Before Solving an Equation**

Solve the equation. Check your solution.

1. $n + 7n - 2 = 22$ **2.** $-15 = 5b + 12 - 2b$

3. $2(n - 8) = 22$ **4.** $6(4 - 3k) = 42$

5. $8g - (5 - 2g) = 15$ **6.** $\dfrac{2}{5}(b - 6) + 11 = 1$

TEXTBOOK REFERENCES
Developing Concepts 3.4
Lessons 3.4, 3.5

KEY WORDS
- multi-step equation
- inverse operations
- like terms
- distributive property

▶**STUDY TIP**
You may find it best to collect variable terms on the side where the coefficient is greater. In Example 5, for example, move the "$-x$" term to the left side because the coefficient 3 in the "$3x$" term is greater than the coefficient -1 in "$-x$."

COLLECTING VARIABLES ON ONE SIDE OF AN EQUATION

If an equation has variables on both sides, you can use inverse operations to remove the variable from one side of the equation.

Example 5

Solve $3x + 2 = -x - 2$.

▶ *Solution*

$3x + 2 = -x - 2$	Write original equation.
$3x + 2 + x = -x - 2 + x$	Add x to each side.
$4x + 2 = -2$	Combine like terms.
$4x = -4$	Subtract 2 from each side.
$x = -1$	Divide each side by 4.

▶ The solution is -1.

Check ✓ $\quad 3x + 2 = -x - 2$ $\qquad$ Write original equation.

$\qquad 3(-1) + 2 \stackrel{?}{=} -(-1) - 2 \qquad$ Substitute -1 for each x.

$\qquad\qquad -3 + 2 \stackrel{?}{=} 1 - 2 \qquad$ Simplify.

$\qquad\qquad\qquad -1 = -1 ✓ \qquad$ Solution is correct.

Example 6

Solve $3x + 7 = 5(x - 1)$.

▶ *Solution*

$3x + 7 = 5(x - 1)$	Write original equation.
$3x + 7 = 5x - 5$	Use distributive property.
$3x + 7 - 3x = 5x - 5 - 3x$	Subtract $3x$ from each side.
$7 = 2x - 5$	Combine like terms.
$12 = 2x$	Add 5 to each side.
$6 = x$	Divide each side by 2.

▶ The solution is 6. Check this in the original equation.

Checkpoint ✓ *Collect Variables on One Side of the Equation*

Solve the equation. Check your solution.

7. $14z = -26 + z$ $\qquad$ **8.** $7b - 2 = b + 10$ $\qquad$ **9.** $2(n + 9) = n - 2$

Some equations have more than one variable term on the same side. To solve these, combine the like terms that are on the same side. You may need to use the distributive property to remove parentheses before combining like terms.

Example 7

Solve $-36 + 2n = -3n + n - 5n$.

▶ *Solution*

$-36 + 2n = -3n + n - 5n$	Write original equation.
$-36 + 2n = -7n$	Combine like terms on the right side.
$-36 = -9n$	Subtract $2n$ from each side.
$4 = n$	Divide each side by -9.

▶ The solution is 4.

Check ✓ $\quad -36 + 2n = -3n + n - 5n$ $\qquad$ Write original equation.

$\qquad -36 + 2(4) \stackrel{?}{=} -3(4) + (4) - 5(4)$ $\qquad$ Substitute 4 for each n.

$\qquad\qquad -28 = -28$ ✓ $\qquad$ Solution is correct.

Example 8

Solve $2y + 3(y - 9) = -(3y - 18) - y$.

▶ *Solution*

$2y + 3(y - 9) = -(3y - 18) - y$	Write original equation.
$2y + 3y + 3(-9) = (-1)(3y) - (-1)(18) - y$	Use distributive property.
$2y + 3y - 27 = -3y + 18 - y$	Simplify.
$5y - 27 = -4y + 18$	Combine like terms.
$9y - 27 = 18$	Add $4y$ to each side.
$9y = 45$	Add 27 to each side.
$y = 5$	Divide each side by 9.

▶ The solution is 5. Check this in the original equation.

Checkpoint ✓ **Collect Variables on One Side of an Equation**

Solve the equation. Check your solution.

10. $3t - 6 = 5t + 8 - 9t$ $\qquad\qquad$ **11.** $4q - 2(q - 3) = q + 6q - 9$

Exercises Linear Equations in One Variable

Solve the equation. Check your solution.

1. $4x + 1 - x = 19$

2. $4a + 3a - 7 = 21$

3. $1.2x + 2.6x = 4.56$

4. $6 = a + a + 4$

5. $n + 2 - 3n = -8$

6. $3z + z + z = 3.2$

7. $-4.5 = 5n - n - n$

8. $16 = 2(y - 1) - 6$

9. $3n + 4(n - 9) = -78$

10. $4x - 7(x - 9) = 42$

11. $6 - 22(p + 6) = 2$

12. $-9 - (8 - 5t) = 18$

13. $\frac{1}{2}(x + 8) = -5$

14. $18 = -\frac{3}{7}(m - 2)$

Is the given number a solution of the equation?

15. $b + 6 = 3(b - 4), b = 9$

16. $10 - 6m = 2(m - 3), m = -3$

17. $-f + 3f = f + 27, f = 27$

18. $a - 8 - 2 = \frac{1}{2}(a - 2), a = 18$

Solve the equation. Check your solution.

19. $6n = 4n + 20$

20. $3m + 6 = -m - 6$

21. $3d - 8 = -6 + d$

22. $4 - 7m = m + 4$

23. $2y + 5 = -y - 4$

24. $0.3k + 1.4 = 4.2 - 0.1k$

25. $2(y - 5) = 15y - 4y$

26. $4(8 - k) = 2k + 16$

27. $\frac{1}{2}(2h + 4) = 3h - 4$

28. $\frac{1}{5}(x + 8) = \frac{3}{5}x$

29. $-22 - 3d = -2d + d - 3d$

30. $m - 16 = 3m + 18 + 2m$

31. $3(6e - 2) + 4(1 - 5e) = e$

32. $5(3 - 4y) + 14y = 7(2 - 5y)$

33. $2(5 - 4x) + x = 3(3x - 11)$

34. $6(2n - 5) = -3(7 - 3n) + 3$

35. $5.5(3 - 2q) = 2(9 - 7q) + 0.5q$

36. $5x - 1 = (4 - 3x)(-2) + 10$

37. $7(2d - 1) + 5(2 - 3d) = 2d$

38. $4 - 0.6a = 3(1.5a + 0.9) + 7.93$

39. $\frac{3}{4}(8x - 12) + 5x = 7 - (x + 30)$

40. $9 + \frac{2}{3}(6x - 18) = 3x - \frac{1}{5}(40 - 10x)$

Topic Review *Linear Equations and Inequalities*

These exercises will help you check that you can simplify expressions and solve linear equations and inequalities in one variable. If you have any questions about simplifying and solving linear equations and linear inequalities, be sure to get them answered before going on to the next section.

Solve the equation or inequality.

1. $8x - 20 + 2x = 60$ **2.** $3x - 1 \leq -7$ **3.** $19 = x + x - 7$

4. $5x + 6 \leq 1$ **5.** $12 - 2x > 10$ **6.** $3(3x - 8) - 4x = 21$

7. $\dfrac{2}{7}(x + 5) = 8$ **8.** $\dfrac{2}{3}(x - 6) \geq 0$ **9.** $6 \geq \dfrac{1}{4}(3 + x)$

10. $4x - 2x + 3 \leq 3 - 1$ **11.** $12m - 4(3 - m) = 20$

12. $3 - 2q = 3q + 18$ **13.** $3(x - 4) = 8x + 23$

14. $75x - 50 < 55x - 30$ **15.** $2.8x - 8.4 < -1.4x + 8.4$

16. $(x - 6)4 > x + 30$ **17.** $-(4 - x) \geq 2(3 - x)$

18. $2(6x - 6) = 8x - 4x + 52$ **19.** $7n - 5(9 - 3n) = 3(n + 12) - 8n$

20. $-7m + 6 - 3m < 2(m - 3)$ **21.** $8(2x + 1) - 5x < 4x - 13$

22. $\dfrac{4}{5}(10x - 20) + 8x = 15x - 7x$ **23.** $\dfrac{1}{9}(36 - 45q) = \dfrac{2}{3}(5q + 31)$

In Exercises 24 and 25, find and correct the error.

24.
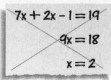

$$7x + 2x - 1 = 19$$
$$9x = 18$$
$$x = 2$$

25.
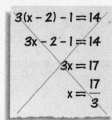

$$3(x - 2) - 1 = 14$$
$$3x - 2 - 1 = 14$$
$$3x = 17$$
$$x = \dfrac{17}{3}$$

In Exercises 26 and 27, match the original inequality with the inequality that is equivalent.

Original inequality Which is equivalent?

26. $5x - 2 < 8 + 6x$ **A.** $x > -10$ **B.** $x > 6$

27. $4(12 - x) + 3x \geq 45$ **A.** $x \geq 3$ **B.** $x \leq 3$

My Review Questions

KEY STANDARD

Algebra 1 **5.0**

Students solve multi-step problems, including word problems, involving linear equations and linear inequalities in one variable and provide justification for each step.

TEXTBOOK REFERENCES
Skills Review pp. 781–782
Lessons 1.6, 3.3–3.9

KEY WORDS
• multi-step equations
• verbal model
• algebraic model

Equations to Solve Problems

WRITING AN ALGEBRAIC MODEL

You may need to write and solve a multi-step linear equation in one variable to solve a problem. The equation is called an **algebraic model**.

To write this model, first use the problem to write a **verbal model** using words and then assign labels to the parts of the verbal model. Use your labels to write the algebraic model. Then use the properties of equality (inverse operations) to solve the equation.

Example 1

A student pays $.86 for three school stickers and a school emblem. An emblem costs $.29. What is the cost for each sticker?

▶ **Solution**

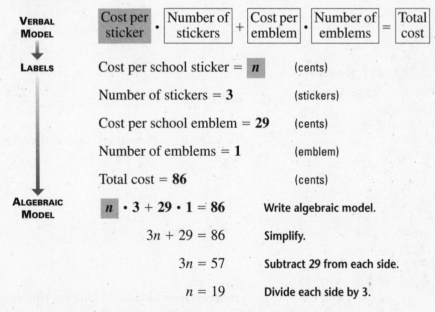

| VERBAL MODEL | Cost per sticker · Number of stickers + Cost per emblem · Number of emblems = Total cost |

LABELS
Cost per school sticker = n (cents)

Number of stickers = **3**   (stickers)

Cost per school emblem = **29**   (cents)

Number of emblems = **1**   (emblem)

Total cost = **86**   (cents)

STUDY TIP
You can use inverse operations to solve an equation.

ALGEBRAIC MODEL

$n \cdot 3 + 29 \cdot 1 = 86$ **Write algebraic model.**

$3n + 29 = 86$ **Simplify.**

$3n = 57$ **Subtract 29 from each side.**

$n = 19$ **Divide each side by 3.**

▶ Each school sticker costs 19 cents, or $.19.

Check ✓ $19 \dfrac{\text{cents}}{\text{sticker}} \cdot 3 \text{ stickers} + 29 \dfrac{\text{cents}}{\text{emblem}} \cdot 1 \text{ emblem} \stackrel{?}{=} 86 \text{ cents}$

$57 \text{ cents} + 29 \text{ cents} \stackrel{?}{=} 86 \text{ cents}$

$86 \text{ cents} = 86 \text{ cents}$ ✓

STUDY TIP
Check solutions of word problems by substituting the values *and their units* in the verbal model.

Example 2

A research company budgeted $1000 each month for an on-line computer information service. The service charges a $250 monthly service fee plus $75 for each hour of use. How many hours can the research company use the service each month and stay within the budget?

▶ **Solution**

Let h = the number of hours used each month.

Then the inequality $250 + 75h \leq 1000$ describes the situation.

$250 + 75h \leq 1000$	**Write original inequality.**
$75h \leq 750$	**Subtract 250 from each side.**
$h \leq 10$	**Divide each side by 75.**

▶ The company can use the service for, at most, 10 hours each month.

Example 3

There are four math tests in the first marking period. Sam's scores on three tests are 85, 91, and 76. What is the lowest score he can get on the fourth test to have an average greater than 85 for the marking period?

▶ **Solution**

You are given three test scores: 85, 91, and 76.

Let x = the fourth test score.

The average of the four scores is $\dfrac{85 + 91 + 76 + x}{4}$.

Translate the information into an inequality.

The average	is greater than	85.
$\dfrac{85 + 91 + 76 + x}{4}$	$>$	85

$\dfrac{85 + 91 + 76 + x}{4} > 85$	**Write original inequality.**
$85 + 91 + 76 + x > 340$	**Multiply each side by 4.**
$252 + x > 340$	**Simplify.**
$x > 88$	**Subtract 252 from each side.**

▶ The fourth test score must be greater than 88. The lowest possible fourth test score is 89.

Example 4

Find the two smallest consecutive integers whose sum is greater than 55.

▶ **Solution**

Let n = the lesser of the two integers.

Then $n + 1$ = the next consecutive integer.

The inequality $n + (n + 1) > 55$ describes the situation.

$n + n + 1 > 55$	Write original inequality.
$2n + 1 > 55$	Combine like terms.
$2n > 54$	Subtract 1 from each side.
$n > 27$	Divide each side by 2.

▶ The two smallest consecutive integers whose sum is greater than 55 are 28 and 29 because they are the smallest integers greater than 27.

Checkpoint ✓ **Inequalities to Solve Problems**

1. You divide a number by -3. Then you subtract 1 from the quotient. The result is at most 5. Find all the numbers that could be solutions.

2. Find the two greatest consecutive odd integers whose sum is less than -11.

Exercises **Inequalities to Solve Problems**

1. Five less than twice a number is at least 13. Find all such numbers.

2. An artist withdrew $14 from a bank in each of the last three weeks and still has more than $65. How much did he start with?

3. Students in math class need an average test score of at least 90 points to earn an A. A student's test scores are 88, 91, and 85. What could the student score on the next test to have an A average?

4. A salesperson earns a salary of $600 per month, plus a commission of 2% of sales. How much must the salesperson sell to have a monthly income of at least $1700?

5. A company's policy is to spend no more than $350,000 a year for salaries. The president's salary is $110,000. The salaries of the eight other employees are equal. How much can be spent on each salary?

SUMMARIZING KEY IDEAS

The steps for solving a multi-step problem involving linear inequalities are similar to the steps for solving a multi-step problem involving linear equations.

TEXTBOOK LINK

On pages S22–S24, you solved multi-step word problems involving linear inequalities in one variable and provided justification. In your textbook, Lesson 6.3, you will continue to use models to solve multi-step word problems involving linear inequalities in one variable.

Topic Review *Multi-Step Problems*

These exercises will help you check that you can solve multi-step problems, including word problems, involving linear equations and linear inequalities in one variable and provide justification for each step. If you have any questions about solving multi-step linear equations and linear inequalities, be sure to get them answered before going on to the next section.

Choose the equation or inequality that best describes the situation. Solve.

1. Frank paid $22.40 for a sweatshirt. It had been discounted by 30%. What was the regular price of the sweatshirt?

    A $x + 0.30x = 22.40$ **B** $x + 30x = 22.40$

    C $x - 0.30x = 22.40$ **D** $x = 22.40(0.30x)$

2. Laura has $16 in her savings account. She earns $4.50 per hour baby-sitting. Laura wants to purchase a sweater for $55. What is the least number of hours Laura must baby-sit in order to buy the sweater?

    A $16x + 4.50 > 55$ **B** $4.50x > 55 + 16$

    C $4.50x + 16 > 55$ **D** $55 > 16 + 4.50x$

My Review Questions

Write an equation or inequality to describe the situation. Solve.

3. Rachel is saving $22 each week from her paycheck. She already has $47 in her account. In how many weeks will her balance be $201?

4. Fifty-three is 19 less than four times a number. Find the number.

5. Find four consecutive integers whose sum is −490.

6. Find four consecutive odd integers whose sum is greater than 48.

7. Four times the sum of a number n and five is equal to 100. Find n.

8. Team A defeated Team B by 13 points. The total number of points scored by both teams was 171. How many points were scored by Team B?

9. Bill paid $53 for a sweater. It was 20% off the regular price. What was the regular price of the sweater?

10. The amount of rainfall over a three-year period was 65 inches, 72 inches, and 59 inches. How many inches of rain must fall during the fourth year for the average rainfall to be at least 68 inches for the four years?

Students graph a linear equation and compute the *x*- and *y*-intercepts (e.g., graph $2x + 6y = 4$). They are also able to sketch the region defined by a linear inequality (e.g., they sketch the region defined by $2x + 6y < 4$).

TEXTBOOK REFERENCES
Developing Concepts 4.2
Lesson 4.2

KEY WORDS
• linear equation
• solution of an equation
• function form

Solutions of Linear Equations

CHECKING SOLUTIONS OF LINEAR EQUATIONS

A **linear equation in two variables** *x* and *y* is an equation that can be written in the form $Ax + By = C$, where *A* and *B* are not both zero.

A **solution of an equation in two variables** is an ordered pair (x, y) that makes the equation true when the values are substituted for the variables in the equation.

Example 1

Determine whether the ordered pair is a solution of $2x - y = 0$.

a. $(-2, -4)$ **b.** $(4, 16)$ **c.** $(5, 5)$

▶ *Solution*

a. $2x - y = 0$ Write original equation.

  $2(-2) - (-4) \stackrel{?}{=} 0$ Substitute −2 for *x* and −4 for *y*.

  $0 = 0$ ✓   Simplify. True statement.

▶ $(-2, -4)$ is a solution of the equation $2x - y = 0$.

b. $2x - y = 0$ Write original equation.

  $2(4) - 16 \stackrel{?}{=} 0$ Substitute 4 for *x* and 16 for *y*.

  $-8 \neq 0$ Simplify. Not a true statement.

▶ $(4, 16)$ is *not* a solution of the equation $2x - y = 0$.

c. $2x - y = 0$ Write original equation.

  $2(5) - 5 \stackrel{?}{=} 0$ Substitute 5 for *x* and 5 for *y*.

  $5 \neq 0$ Simplify. Not a true statement.

▶ $(5, 5)$ is *not* a solution of the equation $2x - y = 0$.

My Notes

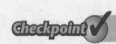 **Check Solutions of Linear Equations**

Determine whether (0, 1) is a solution of the equation.

  1. $x + 2y = 2$ **2.** $3x - y = -1$ **3.** $x = y + 1$

FINDING SOLUTIONS OF LINEAR EQUATIONS

A two-variable equation is written in **function form** if one of its variables is isolated on one side of the equation. For example, $y = 2x + 3$ is in function form while $3x + 2y = 5$ is *not* in function form.

Example 2

Find four ordered pairs that are solutions of $y - 2x = 3$.

▶ *Solution*

Rewrite the equation in function form to make it easier to substitute values into the equation.

$$y - 2x = 3 \qquad \text{Write original equation.}$$
$$y = 2x + 3 \qquad \text{Add 2x to each side.}$$

Choose a few values of x and substitute them into the equation to find the corresponding y-values.

x	$2x + 3$	y
-2	$2(-2) + 3 = -1$	-1
-1	$2(-1) + 3 = 1$	1
0	$2(0) + 3 = 3$	3
1	$2(1) + 3 = 5$	5

▶ $(-2, -1)$, $(-1, 1)$, $(0, 3)$, and $(1, 5)$ are four solutions of $y - 2x = 3$.

Checkpoint ✔ **Find Solutions** *Linear Equations*

Find three ordered pairs that *are* **solutions of the equation.**

4. $3x = 4y - 24$ **5.** $+ y = 3$ **6.** $3x - y = -2$

Exercises *Solutions of Linear Equations*

Determine which or **pair listed is a solution of the equation.**

1. $2x + y = 3$; $(1, \quad)$ **2.** $-3x + y = 5$; $(2, 1), (-2, -1)$

3. $x - 3y = 6$; $, (6, -1)$ **4.** $y = \frac{1}{2}x - 4$; $(-2, -5), (1, -3)$

Find three o **pairs that are solutions of the equation.**

5. $3x +$ **6.** $x + 2y = 2$ **7.** $2y - 3x = 12$

STUDY TIP
You can find a solution of an equation in two variables by choosing a value for one variable and using it to find the value of the other variable.

SUMMARIZING KEY IDEAS
You can decide if an ordered pair is a solution of a linear equation in two variables by substituting the values for the variables. If the result is a true equation, then the ordered pair is a solution of the equation.

TEXTBOOK LINK
On pages S26–S27, you determined which ordered pairs are solutions of an equation and found ordered pairs that are solutions of equations. In your textbook, Lesson 4.2, you will learn more about these topics.

TEXTBOOK REFERENCES
Developing Concepts 4.2
Lessons 4.2, 4.3

KEY WORDS
- linear equation
- solution of an equation
- function form
- graph of an equation

> **STUDY TIP**
> Try to choose values of *x* that include negative values, zero, and positive values to see how the graph behaves to the left and right of the *y*-axis.

✏ **My Notes**

Graphs of Linear Equations

GRAPHING A LINEAR EQUATION

The **graph of an equation** in x and y is the set of *all* points (x, y) that are solutions of the equation. The graph of a linear equation is a straight line.

Example 1

Use a table of values to graph $3x + 4y = 12$.

▶ **Solution**

Rewrite the equation in function form by solving for y.

$$3x + 4y = 12 \qquad \text{Write original equation.}$$

$$4y = -3x + 12 \qquad \text{Subtract } 3x \text{ from each side.}$$

$$\frac{4}{4}y = \frac{-3}{4}x + \frac{12}{4} \qquad \text{Divide each side by 4.}$$

$$y = \frac{-3}{4}x + 3 \qquad \text{Simplify.}$$

Choose a few values of x and make a table of values.

x	$\dfrac{-3}{4}x + 3$	y
-4	$\dfrac{-3}{4}(-4) + 3 = 6$	6
0	$\dfrac{-3}{4}(0) + 3 = 3$	3
4	$\dfrac{-3}{4}(4) + 3 = 0$	0

With this table of values you have found three solutions:

$(-4, 6)$, $(0, 3)$, and $(4, 0)$

Plot the points.

▶ The line through the points is the graph of the equation $3x + 4y = 12$, as shown.

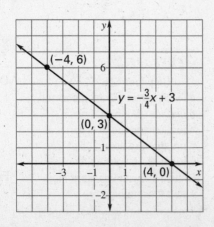

Checkpoint ✓ **Graph Linear Equations**

Use a table of values to graph the equation.

1. $y = 3x - 1$ **2.** $2x + 3y = 12$ **3.** $4x - 5y = 20$

GRAPHING HORIZONTAL AND VERTICAL LINES

TEXTBOOK REFERENCES

Lessons 4.3, 4.4

KEY WORDS
• horizontal line
• vertical line
• *x*-value
• *y*-value

The graph of a line with an equation of the form $y = b$ is a **horizontal line**. The graph of a line with an equation of the form $x = a$ is a **vertical line**.

Example 2

Graph the equation $y = 3$.

▶ **Solution**

The equation does not have x as a variable. The y-value is always 3, regardless of the value of x. For instance, here are some points that are solutions of the equation:

$(-3, 3)$, $(0, 3)$, and $(3, 3)$

▶ The graph of the equation $y = 3$ is a horizontal line 3 units *above* the x-axis.

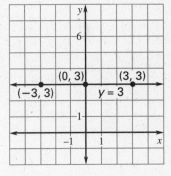

My Notes

Example 3

Graph the equation $x = -4$.

▶ **Solution**

The equation does not have y as a variable. The x-value is always -4, regardless of the value of y. For instance, here are some points that are solutions of the equation:

$(-4, -2)$, $(-4, 0)$, and $(-4, 3)$

▶ The graph of the equation $x = -4$ is a vertical line 4 units *to the left* of the y-axis.

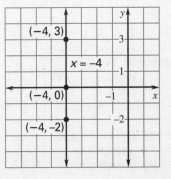

Checkpoint ✔ **Graph Horizontal and Vertical Lines**

Graph the equation.

4. $x = 5$ **5.** $y = -2$ **6.** $y = 0$

USING INTERCEPTS TO MAKE A QUICK GRAPH

TEXTBOOK REFERENCES

Lesson 4.4

KEY WORDS

- *x*-coordinate
- *y*-coordinate
- *x*-intercept
- *y*-intercept

The *x*- and *y*-intercepts of a line provide a quick way to sketch the line. An **x-intercept** of a graph is the *x*-coordinate of a point where the graph crosses the *x*-axis. A **y-intercept** of a graph is the *y*-coordinate of a point where the graph crosses the *y*-axis.

Example 2

Graph the equation $3.5x + 7y = 14$.

▶ **Solution**

Find the intercepts. Substitute 0 for *y* to find the *x*-intercept. Then substitute 0 for *x* to find the *y*-intercept.

$3.5x + 7y = 14$	Write original equation.
$3.5x + 7(0) = 14$	Substitute 0 for *y*.
$3.5x = 14$	Simplify.
$x = 4$	Divide each side by 3.5.

▶ The *x*-intercept is 4. The line crosses the *x*-axis at (4, 0).

$3.5x + 7y = 14$	Write original equation.
$3.5(0) + 7y = 14$	Substitute 0 for *x*.
$7y = 14$	Simplify.
$y = 2$	Divide each side by 7.

▶ The *y*-intercept is 2. The line crosses the *y*-axis at (0, 2).

Draw a coordinate plane that includes the points (4, 0) and (0, 2).

Plot the points (4, 0) and (0, 2) and draw a line through them.

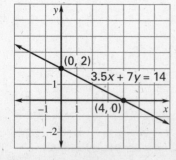

▶**STUDY TIP**

When you make a quick graph, find the intercepts *before* you draw the coordinate plane. This will help you find an appropriate scale on each axis.

SUMMARIZING KEY IDEAS

You can graph a linear equation by using a table of values or by finding the x- and y-intercepts. Using either method, find two or more ordered pairs that are solutions of the equation, plot the points with those ordered pairs, and draw a line through the points.

Checkpoint ✓ **Use Intercepts to Make a Quick Graph**

Find the x-intercept and the y-intercept of the graph of the equation.

7. $-x - 5y = 12$ **8.** $-13x - y = 39$ **9.** $9x - 4y = 54$

10. $-x + 8y = 40$ **11.** $6x + 9y = -81$ **12.** $2x - 17y = -51$

Graph the equation using intercepts.

13. $3y + 2x = 6$ **14.** $4y - x = 8$ **15.** $2x - 5y = 10$

Exercises Graphs of Linear Equations

Use a table of values to graph the equation.

1. $y = 2x - 1$ **2.** $y = 3x + 6$ **3.** $y = 2x + 1$

4. $4x + 2y = 1$ **5.** $y = -4x + 5$ **6.** $y = \frac{1}{2}x + 4$

Graph the equation.

7. $y = -6$ **8.** $x = 4$ **9.** $y = 7$

10. $x = -2$ **11.** $x = -12$ **12.** $y = -9$

Find the x-intercept of the graph of the equation.

13. $x - 2y = 4$ **14.** $x + 4y = -2$ **15.** $2x - 3y = 6$

16. $5x + 6y = 95$ **17.** $-6x - 4y = 42$ **18.** $x + 5y = 10$

Find the y-intercept of the graph of the equation.

19. $x + 5y = 10$ **20.** $y = 4x - 2$ **21.** $y = -3x + 7$

22. $y = 13x + 26$ **23.** $y = 6x - 24$ **24.** $3x - 4y = 16$

Graph the equation using intercepts.

25. $y = 2 - x$ **26.** $3x + 5y = 15$ **27.** $-9x + 2y = 36$

28. $x - 6y = 36$ **29.** $7x - 5y = 35$ **30.** $y = x + 3$

TEXTBOOK LINK

📖 On pages S28–S31, you made a table of values and then graphed equations, graphed vertical and horizontal lines, found the x-intercept and y-intercept of a line, and used the intercepts to sketch a quick graph. In your textbook, Lesson 4.2, you will learn more about these topics.

TEXTBOOK REFERENCES
Lesson 6.8

KEY WORDS
- linear inequality in two variables
- half-plane
- boundary
- graph of a linear inequality in two variables
- solution of a linear inequality in two variables

Graphs of Linear Inequalities

CHECKING SOLUTIONS OF LINEAR EQUATIONS

The graph of a linear equation separates the coordinate plane into three sets of points: points on the line, points above the line, and points below the line. The regions above and below the line are called **half-planes**. The line is called the **boundary** of each half-plane.

The **graph of a linear inequality in two variables** x and y includes all points (x, y) on one side of the boundary line and may also include the points on the boundary line.

A **solution of a linear inequality in two variables** is a point included on the graph of the inequality. A solution will be an ordered pair that makes the inequality true when the values of x and y are substituted into the inequality.

Example 3

Decide which of the given points is a solution of the inequality.

a. $y > x + 2$; $(-5, -1)$ or $(0, 0)$?   **b.** $x < 3$; $(-2, 4)$ or $(3, 2)$?

▶ **Solution**

> **STUDY TIP**
> A *linear inequality* relates two linear expressions with an inequality sign.

a. Try $(-5, 1)$.

$$y > x + 2$$
$$-1 \overset{?}{>} -5 + 2$$
$$-1 > -3 \checkmark$$

▶ $(-5, 1)$ is a solution.

Try $(0, 0)$.

$$y > x + 2$$
$$0 \overset{?}{>} 0 + 2$$
$$0 \not> 2$$

▶ $(0, 0)$ is *not* a solution.

b. Try $(-2, 4)$.

$$x < 3$$
$$-2 < 3 \checkmark$$

▶ $(-2, 4)$ is a solution.

Try $(3, 2)$.

$$x < 3$$
$$3 \not< 3$$

▶ $(3, 2)$ is *not* a solution.

Checkpoint ✓ *Check Solutions of a Linear Inequality*

Decide whether the ordered pair is a solution of the inequality.

1. $y < x + 1$; $(1, -1)$ **2.** $y \leq 2x + 4$; $(3, 2)$ **3.** $y < 2x$; $(-3, -6)$

KEY WORDS
• open half-plane
• closed half-plane

SKETCHING THE GRAPH OF A LINEAR INEQUALITY

The graphs of linear inequalities are shown by shading. If the inequality uses < or >, the boundary line is *not* part of the graph and is drawn as a dashed line. The shaded area is called an **open half-plane**.

If the inequality uses ≤ or ≥, the boundary line *is* part of the graph and is drawn as a solid line. The line and the shaded area together are called a **closed half-plane**.

Example 2

Write the inequality whose graph is shown. Is it an *open* or a *closed* half-plane? (The equation of the boundary line is given.)

a. 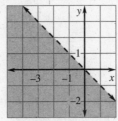   **b.** 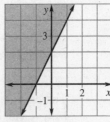   **c.**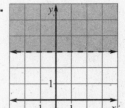

boundary line: boundary line: boundary line:
$y = -x$ $y = 2x + 2$ $y = 3$

▶ **Solution**

a. $y < -x$;   **b.** $y \geq 2x + 2$;   **c.** $y > 3$;
open half-plane   closed half-plane   open half-plane

Checkpoint ✓ **Read a Graph of a Linear Inequality**

Write the inequality whose graph is shown. Is it an *open* or a *closed* half-plane? (The equation of the boundary line is given.)

4. 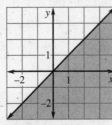   **5.** 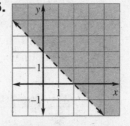   **6.**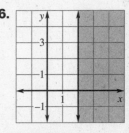

boundary line: boundary line: boundary line:
$y = x$ $y = -x + 2$ $x = 2$

My Notes

Example 3

Sketch the graph of y < 2x + 1.

▶ **Solution**

Sketch a graph of $y = 2x + 1$ as a dashed line because the inequality symbol is <.

Test a point in each half-plane.

Try $(0, 0)$. Try $(-1, 2)$.

$0 \overset{?}{<} 2(0) + 1$ $2 \overset{?}{<} 2(-1) + 1$

$0 < 1 ✓$ $2 \not< -1$

Shade the half-plane containing the point $(0, 0)$.

▶ So the graph of $y < 2x + 1$ is all points *below* the line $y = 2x + 1$.

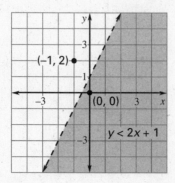

························

The boundary line for a linear inequality involving only x or only y will be a vertical or horizontal line.

Example 4

Sketch the graph of x ≤ −3.

▶ **Solution**

Sketch the vertical line given by $x = -3$. Use a solid line because the inequality symbol is ≤.

Test a point. The origin $(0, 0)$ is *not* a solution because $0 > -3$. Notice that the origin lies to the right of the line.

Shade the half-plane to the *left* of the line.

▶ So the graph of $x \le -3$ is all points to the *left* of the line $x = -3$ and all points *on* the line.

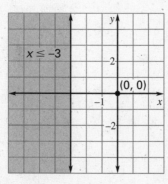

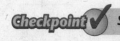 **Checkpoint** ✓ *Sketch the Graph of a Linear Inequality*

Graph the inequality.

7. $y \ge x + 5$ **8.** $y < x$ **9.** $x \ge 5$ **10.** $y < -2$

SUMMARIZING KEY IDEAS

A solution of a linear inequality is an ordered pair that makes the inequality true.

To graph a linear inequality, graph the corresponding equation. Determine if the line should be solid or dashed and whether the solutions are in the half-plane above or below the line. Shade the appropriate region.

Exercises *Graphs of Linear Inequalities*

Decide which of the given points is a solution of the inequality.

1. $y > x + 5$; $(3, 1)$ or $(-6, 0)$?   **2.** $y > x + 1$; $(3, 1)$ or $(0, 2)$?

3. $y + 2 > x + 1$; $(3, 5)$ or $(2, 1)$?   **4.** $y + 3 \geq x + 4$; $(1, 2)$ or $(3, 0)$?

Write the inequality whose graph is shown. Is it an *open* or a *closed* half-plane? (The equation of the boundary line is given.)

5.
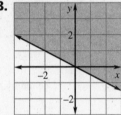
boundary line:
$y = -x - 1$

6.
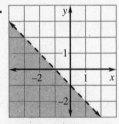
boundary line:
$y = -3x + 3$

7.
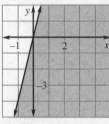
boundary line:
$y = 4x$

8.
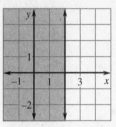
boundary line:
$y = -\dfrac{1}{2}x$

9.
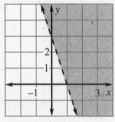
boundary line:
$x = 2$

10.
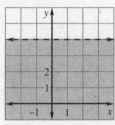
boundary line:
$y = 4$

Graph the inequality.

11. $y > 2x + 5$ **12.** $y \geq 1$ **13.** $y \geq -4$

14. $y < 3x - 1$ **15.** $y < 2x - 3$ **16.** $y + 2 \leq x - 2$

TEXTBOOK LINK

On pages S32–S35, you identified points that are solutions of a given inequality, wrote the inequality from a graph, and graphed inequalities. In your textbook, Lesson 6.8, you will check solutions of linear inequalities and graph linear inequalities.

Topic Review *Graphs of Linear Equations and Linear Inequalities*

These exercises will help you check that you can read and draw graphs of linear equations and linear inequalities. If you have any questions about graphing equations and inequalities, be sure to get them answered before going on to the next section.

Decide which of the given points is a solution of the equation or inequality.

1. $3y - 5x = 6$; $(0, 1)$ or $(3, 7)$?   **2.** $y \le x + 5$; $(-3, 2)$ or $(0, 8)$?

3. $y > 4x - 3$; $(0, 0)$ or $(2, 2)$?   **4.** $2x + y = 12$; $(6, 3)$ or $(3, 6)$?

Write the inequality whose graph is shown. Is it an *open* or a *closed* half-plane? (The equation of the boundary line is given.)

5.
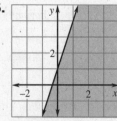
boundary line:
$y = 3x + 1$

6.

boundary line:
$y = -4$

7.

boundary line:
$y = -x + 2$

Find three ordered pairs that are solutions of the equation or inequality.

8. $2x - y = 2$ **9.** $y > x + 3$ **10.** $y \le 2x - 4$

11. $5x + 3y = 1$ **12.** $y < 6x$ **13.** $4y - 7x = -8$

Find the *x*-intercept and *y*-intercept of the graph of the equation.

14. $y = x - 5$ **15.** $3x + 2y = 9$ **16.** $5y - 4x = 20$

Graph the equation or inequality.

17. $y = 4$ **18.** $y > x + 2$ **19.** $y = x - 6$

20. $y = 6x - 3$ **21.** $x = -3$ **22.** $8x - 2y = 4$

23. $y \le 3x - 2$ **24.** $x < 2$ **25.** $6y + 12x = 12$

26. $y \ge x$ **27.** $y < -2x + 1$ **28.** $5y - x = 5$

My Review Questions

Algebra 1 **7.0** Students verify that a point lies on a line, given an equation of the line. Students are able to derive linear equations by using the point-slope formula.

TEXTBOOK REFERENCES
Developing Concepts 4.5
Lessons 4.2, 4.5

KEY WORDS
• rise
• run
• slope
• collinear points

Slope of a Line

FINDING THE SLOPE OF A LINE

You can calculate the slope of a line just from knowing two points on the line. The **rise** is the difference of the y-coordinates and the **run** is the difference of the x-coordinates.

> For any two points (x_1, y_1) and (x_2, y_2) on a line, the slope m of the line is given by the formula:
> $$m = \frac{\text{rise}}{\text{run}} = \frac{\text{vertical change}}{\text{horizontal change}} = \frac{y_2 - y_1}{x_2 - x_1}$$

▶**STUDY TIP**
Either point can be assigned the coordinates (x_1, y_1) or (x_2, y_2). Just be sure to subtract the y-values and the x-values in the same order.

Example 1

Find the slope of a line that passes through the points (4, 2) and (5, 4).

▶ **Solution**

Substitute in the formula $m = \dfrac{y_2 - y_1}{x_2 - x_1}$.

$$m = \frac{y_2 - y_1}{x_2 - x_1} = \frac{4 - 2}{5 - 4} = \frac{2}{1} = 2$$

▶ The slope is 2. The line rises 2 units for every 1-unit change to the right.

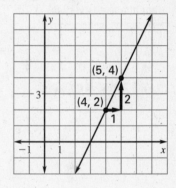

▶**STUDY TIP**
The slope of a line that rises from left to right is positive. The slope of a line that falls from left to right is negative.

Example 2

Find the slope of a line that passes through the points (−3, 1) and (2, −2).

▶ **Solution**

Substitute in the formula $m = \dfrac{y_2 - y_1}{x_2 - x_1}$.

$$m = \frac{y_2 - y_1}{x_2 - x_1} = \frac{-2 - 1}{2 - (-3)} = \frac{-3}{5} = -\frac{3}{5}$$

▶ The slope is $-\dfrac{3}{5}$. The line falls 3 units for every 5-unit change to the right.

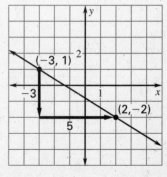

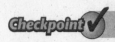 **Find the Slope of a Line**

Find the slope of the line that passes through the two points.

1. $(6, 7), (1, 3)$ **2.** $(-1, 3), (2, 4)$ **3.** $(-2, 3), (2, 1)$

Example 3

Find the slope of the line described. Graph the line.

a. A line that passes through the points $(0, 2)$ and $(5, 2)$

b. A line that passes through the points $(-3, 4)$ and $(-3, 2)$

> **▶ STUDY TIP**
> If the slope of a line is 0, then the line is horizontal. If the slope of a line is undefined, then the line is vertical.

▶ Solution

a. $m = \dfrac{y_2 - y_1}{x_2 - x_1}$

$= \dfrac{2 - 2}{5 - 0}$

$= \dfrac{0}{5}$, or 0

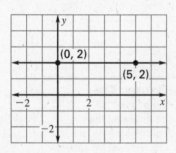

▶ The slope of the line is zero. The line is horizontal.

▶ Solution

b. $m = \dfrac{y_2 - y_1}{x_2 - x_1}$

$= \dfrac{4 - 2}{-3 - (-3)}$

$= \dfrac{2}{0}$, or undefined

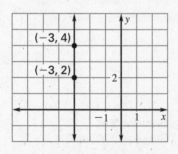

▶ The slope of the line is undefined. The line is vertical.

My Notes

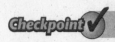 **Find the Slope of a Line**

Find the slope of a line that passes through the two points. State whether it is a *horizontal* or a *vertical* line.

4. $(1, 3), (2, 3)$ **5.** $(0, 5), (3, 5)$ **6.** $(-5, 2), (-5, 1)$

DETERMINING WHETHER POINTS ARE COLLINEAR

Collinear points are points that lie on the same line. The slope between any pair of points on a line will be the same as the slope between any other pair of points on the same line.

Example 4

Determine whether $(-4, -1)$, $(-2, 1)$, and $(2, 5)$ are collinear.

▶ **Solution**

SUMMARIZING KEY IDEAS

One characteristic of a line is its slope, a measure of its steepness. Choose two points along the line. The run is the difference in the *x*-coordinates. The rise is the difference in the *y*-coordinates. The slope is the ratio $\dfrac{\text{rise}}{\text{run}}$.

Determine the slopes using each pair of points.

(−4, −1) and (−2, 1):

$$m = \frac{1 - (-1)}{-2 - (-4)}$$

$$= \frac{2}{2} = 1$$

(−2, 1) and (2, 5):

$$m = \frac{5 - 1}{2 - (-2)}$$

$$= \frac{4}{4} = 1$$

(−4, −1) and (2, 5):

$$m = \frac{5 - (-1)}{2 - (-4)}$$

$$= \frac{6}{6} = 1$$

▶ The slopes are all equal, so the points are collinear.

Checkpoint ✓ *Determine Whether Points are Collinear*

The coordinates of three points are given. Determine whether the points are collinear. Check your answer by graphing the points.

7. $(3, 1)$, $(1, -2)$, $(5, 4)$ **8.** $(2, -6)$, $(0, -2)$, $(6, 10)$

Exercises *Slope of a Line*

TEXTBOOK LINK

On pages S37–S39, you found the slope of a line by using pairs of coordinate points on the line. You also found the slopes of horizontal and vertical lines. Then you determined whether points are collinear. In your textbook, Lesson 4.5, you will use the slope formula to determine whether the slope of a line is positive, negative, zero, or undefined.

Find the slope of a line that passes through the two points.

1. $(4, 7)$, $(7, 11)$ **2.** $(-4, 4)$, $(2, -5)$ **3.** $(-3, 1)$, $(3, -4)$

4. $(6, -3)$, $(1, 2)$ **5.** $(7, -1)$, $(2, 3)$ **6.** $(5, -2)$, $(4, 3)$

Find the slope of a line that passes through the two points. State whether it is a *horizontal* or a *vertical* line.

7. $(-3, 0)$, $(-3, 2)$ **8.** $(4, 7)$, $(3, 7)$ **9.** $(2, 1)$, $(0, 1)$

The coordinates of three points are given. Determine whether the points are collinear. Check your answer by graphing the points.

10. $(-2, 1)$, $(-2, 4)$, $(2, 4)$ **11.** $(-2, 1)$, $(0, 4)$, $(2, 7)$

12. $(1, -2)$, $(-1, -5)$, $(5, 4)$ **13.** $(-3, 4)$, $(0, 2)$, $(-3, 0)$

TEXTBOOK REFERENCES
Developing Concepts 4.7
Lessons 4.7, 5.1

KEY WORDS
• slope
• *y*-intercept
• slope-intercept form

Slope-Intercept Form

FINDING THE SLOPE AND *y*-INTERCEPT

The equation $y = mx + b$ is called the **slope-intercept form** of a line. The graph of the equation is a line with slope m and *y*-intercept b.

Example 1

Find the slope and the *y*-intercept of the graph of 4*x* + 3*y* = 12.

▶ **Solution**

Rewrite the equation in slope-intercept form.

$4x + 3y = 12$ **Write original equation.**

$3y = -4x + 12$ **Subtract 4*x* from each side.**

$y = -\dfrac{4}{3}x + 4$ **Divide each side by 3. $m = -\dfrac{4}{3}$ and $b = 4$.**

▶ The slope is $-\dfrac{4}{3}$. The *y*-intercept is 4.

My Notes

Checkpoint ✓ *Find the Slope and y-Intercept*

Find the slope and the *y*-intercept of the graph of the equation.

1. $y = -\dfrac{2}{3}x$ **2.** $x + 5y = 10$ **3.** $8x - y = 2$

Example 2

Write in slope-intercept form an equation of the line whose slope is −3 and whose *y*-intercept is 0.

▶ **Solution**

The values for the slope and *y*-intercept are $m = -3$ and $b = 0$.

$y = mx + b$ **Write the slope-intercept form.**

$y = -3x + 0$ **Substitute −3 for *m* and 0 for *b*.**

$y = -3x$ **Simplify.**

▶ An equation of the line is $y = -3x$.

Checkpoint ✓ *Write an Equation of a Line*

Write in slope-intercept form an equation of the line whose slope is *m* and whose *y*-intercept is *b*.

4. $m = -2,$
$b = 0$

5. $m = -5,$
$b = 2$

6. $m = 1,$
$b = 3$

GRAPHING USING THE SLOPE-INTERCEPT FORM

Although linear equations can be graphed using a table of values, a more efficient method uses only the slope and the *y*-intercept.

Example 3

Graph the equation *x* − 2*y* = 4.

▶ *Solution*

❶ Write the equation in the form $y = mx + b$.

$x - 2y = 4$ .... Write original equation.

$-2y = -x + 4$ .... Subtract *x* from each side.

$y = \dfrac{1}{2}x - 2$ .... Divide each side by −2.

▶ Then $m = \dfrac{1}{2}$ and $b = -2$.

❷ The *y*-intercept is −2. Plot the point whose coordinates are $(0, -2)$.

❸ From that point, use the slope $\dfrac{1}{2}$ to locate another point. Move 2 units to the right and 1 unit up.

$$\text{slope} = \dfrac{1}{2} \begin{matrix} \leftarrow \textbf{up 1} \\ \leftarrow \textbf{right 2} \end{matrix}$$

❹ Draw a line through the two points.

▶ The graph of $x - 2y = 4$ is shown at the right.

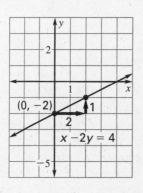

My Notes

Checkpoint ✔ **Graph an Equation Using Slope-Intercept Form**

Graph the equation using the slope and the *y*-intercept.

7. $y = -\dfrac{4}{3}x + 2$ **8.** $y - 2x = 0$ **9.** $5x + 2y = -10$

Remember, $y = mx + b$ is the slope-intercept form of the equation of a line. When an equation is in this form, the slope of the line is given by m and the *y*-intercept by b, so the line crosses the *y*-axis at $(0, b)$.

Exercises   *Slope-Intercept Form*

Find the slope and the *y*-intercept of the graph of the equation.

1. $y = 3x + 6$ **2.** $5x + 3y = 15$ **3.** $x - 3y = 9$

4. $x - 4y = 20$ **5.** $y = -2x$ **6.** $3x + 2y = 6$

Write in slope-intercept form an equation of the line with the given slope and *y*-intercept.

7. $m = 4$,   **8.** $m = \dfrac{2}{3}$,   **9.** $m = \dfrac{3}{4}$,   **10.** $m = 0$,

   $b = 2$ $b = 4$ $b = -3$ $b = -1$

Use the graph to find the slope and *y*-intercept of the line. Then write an equation of the line in slope-intercept form.

11. 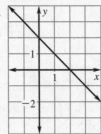   **12.** 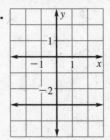   **13.**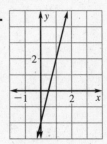

Graph the equation using the slope and the *y*-intercept.

14. $x - 3y = 9$ **15.** $x - 6y = 18$ **16.** $x + 2y = 8$

17. $2x + 4y = 20$ **18.** $3x - 2y = 4$ **19.** $4x - y = 2$

20. $3x = 4y$ **21.** $3x - 4y = 8$ **22.** $-2x - y = 5$

SUMMARIZING KEY IDEAS

If you know the slope and the *y*-intercept of a line, you can write an equation of the line. For example, if the slope of a line is 1 and the *y*-intercept is −3, substitute 1 for *m* and −3 for *b* into the slope-intercept form for the equation of a line. Then an equation of the line is $y = x - 3$.

TEXTBOOK LINK

On pages S40–S42, you found the slope and the *y*-intercept of a line, and graphed an equation in slope-intercept form. In your textbook, Lesson 4.7, you will graph an equation in slope-intercept form, use a linear model, and identify parallel lines.

TEXTBOOK REFERENCES
Developing Concepts 5.2
Lessons 5.2, 5.3, and 5.4

KEY WORDS
- slope
- y-intercept
- slope-intercept form of an equation
- point-slope form of an equation
- standard form of an equation
- coefficient

Writing Equations

WRITING EQUATIONS GIVEN POINT AND SLOPE

You can write an equation of a line in a number of different forms. You have been using **slope-intercept form** $y = mx + b$, where m is the slope and b the y-intercept.

The **point-slope form** of an equation with slope m that passes through the point (x_1, y_1) is $y - y_1 = m(x - x_1)$.

An equation is written in **standard form** when it is in the form $Ax + By = C$, where A and B are not both zero.

Example 1

Write in standard form an equation of a line that has a slope of $\frac{2}{3}$ and passes through the point (3, 6). Use integer coefficients.

▶ **Solution**

❶ Write an equation in point-slope form.

$y - y_1 = m(x - x_1)$ Write point-slope form.

$y - 6 = \frac{2}{3}(x - 3)$ Substitute $\frac{2}{3}$ for m, 6 for y_1, and 3 for x_1.

❷ Change the equation to standard form. Use integer coefficients.

$y - 6 = \frac{2}{3}(x - 3)$ Write equation in point-slope form.

$3(y - 6) = 3\left[\frac{2}{3}(x - 3)\right]$ Multiply each side by 3.

$3(y - 6) = 2(x - 3)$ Simplify.

$3y - 18 = 2x - 6$ Use distributive property.

$3y = 2x + 12$ Add 18 to each side.

$-2x + 3y = 12$ Subtract 2x from each side.

▶ An equation of the line in standard form is $-2x + 3y = 12$.

My Notes

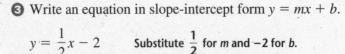

Checkpoint ✓ *Write an Equation from a Point and the Slope*

Write in standard form an equation of the line that passes through the given point and has the given slope. Use integer coefficients.

1. $(-2, -4), m = 3$ **2.** $(-4, -1), m = 1$ **3.** $(1, 1), m = \dfrac{3}{4}$

WRITING AN EQUATION GIVEN TWO POINTS

When you can determine the slope of a line and its y-intercept or another point on the line from its graph, you can write an equation of the line.

Example 2

Write in slope-intercept form an equation of the line shown.

▶ **Solution**

❶ The line intersects the y-axis at the point whose coordinates are $(0, -2)$. The y-intercept is -2.

 My Notes

❷ Use the graph to find another point on the line, such as $(2, -1)$ or $(-2, -3)$. Use the points to find the slope.

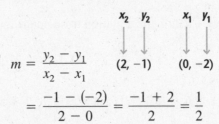

$$m = \frac{y_2 - y_1}{x_2 - x_1} \qquad (2, -1) \qquad (0, -2)$$

$$= \frac{-1 - (-2)}{2 - 0} = \frac{-1 + 2}{2} = \frac{1}{2}$$

The slope is $\dfrac{1}{2}$.

❸ Write an equation in slope-intercept form $y = mx + b$.

$$y = \frac{1}{2}x - 2 \qquad \text{Substitute } \frac{1}{2} \text{ for } m \text{ and } -2 \text{ for } b.$$

▶ An equation the line in slope-intercept form is $y = \dfrac{1}{2}x - 2$.

Checkpoint ✓ *Write an Equation of a Line*

Write in slope-intercept form an equation of the line shown. Then write the equation in standard form. Use integer coefficients.

4.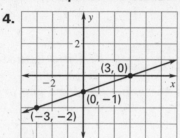

5.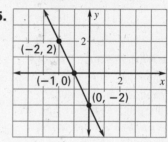

Example 3

Write in standard form an equation of the line that passes through $(-3, -12)$ and $(-11, 12)$. Use integer coefficients.

▶ *Solution*

To write an equation in standard form from two points on the line, you need to first write an equation in another form and then change it to standard form.

❶ Find the slope of the line through the two points.

$$m = \frac{y_2 - y_1}{x_2 - x_1} = \frac{12 - (-12)}{-11 - (-3)} = \frac{12 + 12}{-11 + 3} = \frac{24}{-8} = -3$$

The slope is -3.

❷ Write an equation of the line in point-slope form.

$y - y_1 = m(x - x_1)$

$y - 12 = -3(x - (-11))$ $m = -3; (x_1, y_1) = (-11, 12)$

$y - 12 = -3(x + 11)$

❸ Change the equation of the line to standard form.

$y - 12 = -3(x + 11)$	Write point-slope form.
$y - 12 = -3x + (-3)(11)$	Use distributive property.
$y - 12 = -3x - 33$	Simplify.
$y = -3x - 21$	Add 12 to each side.
$3x + y = -21$	Add 3x to each side.

▶ An equation of the line in standard form is $3x + y = -21$.

My Notes

Checkpoint ✓ **Write an Equation from Two Points**

Write in standard form an equation of the line that passes through the given points. Use integer coefficients.

6. $(-1, -7), (2, 8)$ **7.** $(-2, -9), (1, 8)$ **8.** $(5, 6), (6, 9)$

Exercises **Writing Equations of Lines**

SUMMARIZING KEY IDEAS

To write an equation of a line, you need to know the slope and the y-intercept, the slope and a point on the line, or two points on the line.

Rewrite the equation of the line in standard form.

1. $y = 3x$ **2.** $y = \dfrac{4}{5}x - 8$ **3.** $y - 2 = \dfrac{1}{3}(x + 1)$

Write in slope-intercept form an equation of the line that passes through the given point and has the given slope. Then write the equation in standard form. Use integer coefficients.

4. $(0, 0), m = \dfrac{1}{2}$ **5.** $(2, 6), m = \dfrac{3}{2}$

6. $(-1, 1), m = -2$ **7.** $(-1, 5), m = -3$

8. $(4, -3), m = -\dfrac{5}{4}$ **9.** $(6, 1), m = -\dfrac{1}{5}$

In Exercises 10–15, write in point-slope form an equation of the line that passes through the two points. Then write the equation in standard form. Use integer coefficients.

10. $(-2, -12)$ and $(5, 2)$ **11.** $(-6, 8)$ and $(-9, 16)$

12. $(2, -7)$ and $(-1, -10)$ **13.** $(5, -1)$ and $(-1, 8)$

14. $\left(\dfrac{1}{2}, 0\right)$ and $\left(3, 2\dfrac{1}{2}\right)$ **15.** $\left(\dfrac{1}{3}, 1\right)$ and $\left(2\dfrac{2}{3}, 5\right)$

TEXTBOOK LINK

📖 On pages S43–S46, you wrote an equation of a line using slope and a point on the line. You also wrote equations of lines using a point on the line and the y-intercept and using any two points on the line. In your textbook, Lessons 5.1 and 5.2, you will learn more about writing equations.

16. Write an equation of the line shown at the right.

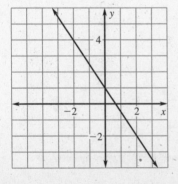

Topic Review *Slope and Equations of Lines*

These exercises will help you check that you can find the slope of a line and write an equation of a line if you know the slope and a point on the line, or if you know two points on the line. If you have any questions about slopes and equations of lines, be sure to get them answered before going on to the next section.

Find the slope of a line that passes through the two points. State whether the line is *horizontal, vertical,* or *neither*.

1. $(1, 3), (5, -3)$ **2.** $(-5, 2), (3, 2)$ **3.** $(-1, 2), (-1, 4)$

4. $(-2, 1), (3, -5)$ **5.** $(-5, 2), (2, -4)$ **6.** $(9, -2), (3, 4)$

7. $(6, -3), (6, 3)$ **8.** $(4, 0), (-4, 0)$ **9.** $(2, 4), (4, -2)$

The coordinates of three points are given. Determine whether the points are collinear. Check your answer by graphing the points.

10. $(-3, 4), (0, 2), (-3, 0)$ **11.** $(3, 3), (-1, -1), (0, 0)$

Find the slope and *y*-intercept of the graph of the equation.

12. $y = -x - 3$ **13.** $x - 2y = 8$ **14.** $3x + 5y = 10$

Graph the equation using the slope and *y*-intercept.

15. $2x - 3y = 6$ **16.** $5x + y = 2$ **17.** $x + 4y = 16$

Write in slope-intercept form an equation of the line shown. Then write the equation in standard form. Use integer coefficients.

18. 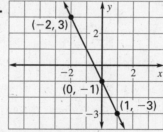   **19.**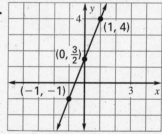

Write in point-slope form an equation of the line that passes through the two points. Then write the equation in standard form. Use integer coefficients.

20. $(6, -7), (8, 1)$ **21.** $(-2, 6), (3, 0)$ **22.** $(0, 5), (5, 0)$

23. $(10, 1), (12, 4)$ **24.** $(-3, -4), (4, 3)$ **25.** $(4, -8), (-8, 4)$

My Review Questions

Algebra 1 **9.0** Students solve a system of two linear equations in two variables algebraically and are able to interpret the answer graphically. Students are able to solve a system of two linear inequalities in two variables and to sketch the solution sets.

TEXTBOOK REFERENCES
Lessons 7.2, 7.3

KEY WORDS
- solution of a system of linear equations
- coefficient

Systems of Linear Equations

SOLVING LINEAR SYSTEMS BY SUBSTITUTION

A **solution of a system of linear equations** is a point that lies on the graph of each equation. The **substitution method** is one technique that can be used to solve a system of linear equations algebraically.

Example 1

Solve the linear system.

$$x - y = 3 \qquad \text{Equation 1}$$
$$2x - 3y = 5.5 \qquad \text{Equation 2}$$

▶ **Solution**

❶ **Solve** one of the equations for one of the variables.

In Equation 1, if $x - y = 3$, then $x = y + 3$.

❷ **Substitute** this expression into the other equation. Solve.

$2x - 3y = 5.5$	Write Equation 2.
$2(y + 3) - 3y = 5.5$	Substitute $y + 3$ for x.
$2y + 6 - 3y = 5.5$	Use distributive property.
$-y + 6 = 5.5$	Combine like terms.
$-y = -0.5$	Subtract 6 from each side.
$y = 0.5$	Divide each side by -1.

> **STUDY TIP**
> The substitution method is convenient to use when the coefficient of one of the variables in the equations in the system is 1 or −1. It is then useful to solve for that variable.

❸ **Substitute** the result in either equation. Solve for the other variable.

$x - y = 3$	Equation 1
$x - 0.5 = 3$	Substitute 0.5 for y.
$x = 3.5$	Add 0.5 to each side.

▶ The solution is (3.5, 0.5).

❹ **Check** ✓ To check the solution, substitute 3.5 for x and 0.5 for y in each of the original equations.

My Notes

Example 2

Solve the linear system.

$$2a - b = -3 \qquad \text{Equation 1}$$
$$4a = 2b - 3 \qquad \text{Equation 2}$$

▶ **Solution**

In Equation 1, if $2a - b = -3$, then $b = 2a + 3$.

$4a = 2b - 3$	Write Equation 2.
$4a = 2(2a + 3) - 3$	Substitute $2a + 3$ for b.
$4a = 4a + 6 - 3$	Use distributive property.
$4a = 4a + 3$	Simplify.
$0 = 3$	Subtract $4a$ from each side.

▶ The result is not possible. So, there is
no solution to the linear system; there
is no ordered pair that is a solution of
both equations. The graph shows that
the lines never intersect.

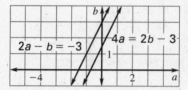

Checkpoint ✔ **Solve by the Substitution Method**

Use the substitution method to solve the linear system. Check the solution in each of the original equations.

1. $y = 2x$
$7x - y = 35$

2. $a + 3b = 5$
$2a - 4b = -5$

3. $3m + 6n = -15$
$m - 3n = 10$

SOLVING SYSTEMS BY ADDITION/SUBTRACTION

Another algebraic method for solving systems of linear equations is the
addition/subtraction method.

❶ **Add** or **subtract** the given equations to eliminate one variable.

❷ **Solve** the resulting equation for the remaining variable.

❸ **Replace** the value of the known variable in one of the original
equations to find the value of the unknown variable.

❹ **Check** the solution in each of the original equations.

▶ **STUDY TIP**
It is convenient to use the addition/subtraction method when the coefficient of one variable in one of the equations is the same as or the opposite of the coefficient of the same variable in the other equation.

▶ **STUDY TIP**
The *standard form of an equation* is $Ax + By = C$, where A and B are not both zero.

Example 3

Solve the linear system.

$$8m + 12n = 32 \qquad \text{Equation 1}$$
$$12n = 3m - 45 \qquad \text{Equation 2}$$

▶ *Solution*

❶
$$8m + 12n = 32$$
$$\underline{-(-3m + 12n = -45)} \qquad \text{Write Equation 2 in standard form.}$$

❷
$$11m \qquad = 77 \qquad \text{Subtract to eliminate the } n\text{-terms.}$$
$$m \qquad = 7 \qquad \text{Divide each side by 11.}$$

❸
$$12n = 3m - 45 \qquad \text{Write either original equation.}$$
$$12n = 3(7) - 45 \qquad \text{Substitute 7 for } m.$$
$$12n = 21 - 45 \qquad \text{Simplify.}$$
$$12n = -24 \qquad \text{Simplify.}$$
$$n = -2 \qquad \text{Divide each side by 12.}$$

▶ The solution of the system is $(7, -2)$. Check by substituting.

Checkpoint ✔ *Use the Addition/Subtraction Method*

Use the addition/subtraction method to solve the linear system. Then check your solution.

4. $x - y = -8$
$x + y = 12$

5. $x + y = 10$
$2x - y = -1$

6. $a + b = 0$
$a - 3b = -8$

SOLVING SYSTEMS BY LINEAR COMBINATIONS

Another algebraic method for solving systems of linear equations is the **linear combination method**.

❶ *Arrange* the equations with like terms in columns.

❷ *Multiply* one or both equations by an appropriate number to obtain new coefficients for x (or for y) that are the same or are opposites.

❸ *Use* the addition/subtraction method to solve the system.

❹ *Check* the solution in each of the original equations.

Example 3

Solve the linear system.

$$4x + 7y = 9 \quad \text{Equation 1}$$
$$9y = -6x + 15 \quad \text{Equation 2}$$

▶ **Solution**

❶ **Write** the equations in standard form and line up the coefficients.

$$4x + 7y = 9 \quad \text{Write Equation 1.}$$
$$6x + 9y = 15 \quad \text{Rearrange Equation 2.}$$

❷ **Multiply** the first equation by 3 and the new second equation by 2. Then the coefficients of the x-terms will each be 12 and the x-terms can be eliminated.

$4x + 7y = 9$	Multiply by 3. →	$12x + 21y = 27$
$6x + 9y = 15$	Multiply by 2.→	$12x + 18y = 30$

$$\begin{array}{ll} 12x + 21y = 27 & \text{Revised Equation 1} \\ -(12x + 18y = 30) & \text{Revised Equation 2} \\ \hline 3y = -3 & \text{Subtract to eliminate the } x\text{-terms.} \\ y = -1 & \text{Solve for } y. \end{array}$$

❸ **Substitute** -1 for y in either original equation. Then solve for x.

$$\begin{array}{ll} 9y = -6x + 15 & \text{Original Equation 2} \\ 9(-1) = -6x + 15 & \text{Substitute } -1 \text{ for } y. \\ -9 = -6x + 15 & \text{Simplify.} \\ -24 = -6x & \text{Subtract 15 from each side.} \\ 4 = x & \text{Divide each side by } -6. \end{array}$$

▶ The solution of the system is $(4, -1)$.

❹ **Check ✓** Check the solution in each of the original equations.

$$\begin{array}{ll} 4x + 7y = 9 & 9y = -6x + 15 \\ 4(4) + 7(-1) \stackrel{?}{=} 9 & 9(-1) \stackrel{?}{=} -6(4) + 15 \\ 16 + (-7) \stackrel{?}{=} 9 & -9 \stackrel{?}{=} -24 + 15 \\ 9 = 9 ✓ & -9 = -9 ✓ \end{array}$$

My Notes

SUMMARIZING KEY IDEAS

You can solve a system of linear equations in two variables by the substitution method, the addition/subtraction method, or the linear combination method.

Checkpoint ✓ **Solve by the Linear Combination Method**

Use the linear combination method to solve the linear system. Then check your solution.

7. $3a - 4b = 1$
$12a - b = -11$

8. $-5c + 3d = -16$
$-10c + d = -22$

9. $8u - 4v = 16$
$4u + 5v = 22$

Exercises Systems of Linear Equations

In Exercises 1–15, use substitution, addition/subtraction, or linear combinations to solve the linear system. Tell which method you chose. Check the solution in each of the original equations.

1. $5 = p + 5q$
$p + q = -3$

2. $3x + 2y = 9$
$x + y = 3$

3. $2p - 5q = 11$
$2p + 5q = 1$

4. $2c + 5d = 44$
$-6c + 5d = 8$

5. $8r - 5s = -11$
$3s = 4r - 11$

6. $9p + 4q = -17$
$12q = -3 - 3p$

7. $3x - 2y = 6$
$5x + 7y = 41$

8. $x + 3y = -4$
$2y + 3x = 2$

9. $4a - 7b = 3$
$16a - 7b = 12$

10. $5m - 2n = 8$
$3m - 5n = 1$

11. $7a - 3b = -9$
$9b - 4a = -24$

12. $-2m - n = 5$
$-2m - 3n = -7$

13. $3y = -5 - 2x$
$7x - 3y = 23$

14. $3x - y = 8$
$4y - 3x = 4$

15. $-3x + 4y = -6$
$-6y + 5x = 8$

16. Use the graph to estimate the solution of the linear system. Then check your solution algebraically.

$6x + 4y = 22$
$2x - 8y = -9$

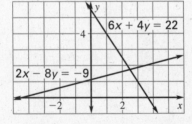

TEXTBOOK LINK

📖 On pages S48–S52, you used the substitution method, the addition/subtraction method, and linear combinations to solve systems of linear equations in two variables. In your textbook, Lessons 7.1, 7.2, and 7.3, you will solve systems of linear equations by graphing, substitution, and linear combination.

17. Show algebraically that there is no solution of the linear system at the right. Graph the lines to show that they never intersect.

$2d = 3e$
$9e - 6d = 9$

TEXTBOOK REFERENCES
Developing Concepts 7.6
Lesson 7.6

KEY WORDS
- linear inequality
- half-plane
- system of linear inequalities

Systems of Linear Inequalities

To solve a system of linear inequalities in two variables x and y, first solve each inequality for y.

GRAPHING A SYSTEM OF LINEAR INEQUALITIES

❶ *Graph* the boundary lines of each inequality. Use a dashed line if the inequality is < or > and a solid line if the inequality is ≤ or ≥.

❷ *Shade* the appropriate half-plane for each inequality.

❸ *Identify* the solution of the system of inequalities as the intersection of the half-planes from Step 2.

Example 1

Graph the system of linear inequalities.

$x + y > 2$ **Inequality 1**

$x - y \geq 3$ **Inequality 2**

▶ *Solution*

Rewrite the inequalities to isolate y.

$$x + y > 2 \qquad\qquad x - y \geq 3$$
$$y > -x + 2 \qquad\qquad -y \geq -x + 3$$
$$y \leq x - 3$$

> **STUDY TIP**
> Reverse the direction of the inequality when you multiply or divide by a negative number.

> **STUDY TIP**
> The points in the solution of a system of inequalities are the only points whose coordinates make all of the inequalities in the system true.

Graph $y > -x + 2$. First graph the boundary line. The graph of $y > -x + 2$ is the half-plane *above* the *dashed* line $y = -x + 2$.

Graph $y \leq x - 3$ in the same coordinate plane. First graph the boundary line. The graph of $y \leq x - 3$ is the half-plane *on and below* the *solid* line $y = x - 3$.

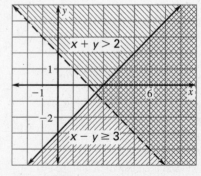

▶ The graph of the solution is the overlap, or *intersection*, of the two half-planes (the double-shaded region). The solution includes points on the boundary line $y = x - 3$, but not on the boundary line $y = -x + 2$.

My Notes

Example 2

Graph of the system of linear inequalities.

$y < 4$ **Inequality 1**

$y > 1$ **Inequality 2**

▶ *Solution*

The graph of Inequality 1 is the half-plane *below* the *dashed* horizontal line $y = 4$.

The graph of Inequality 2 is the half-plane *above* the *dashed* horizontal line $y = 1$.

▶ The graph of the system is the horizontal band that lies between the two dashed horizontal lines.

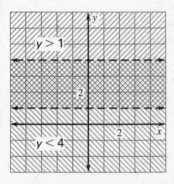

Checkpoint ✔ *Graph a System of Linear Inequalities*

Graph the system of linear inequalities.

1. $y < x + 3$
$y < 5 - 2x$

2. $3x + y \geq 0$
$2x - y \leq -4$

3. $x - 2y < 3$
$2x + y > 8$

Exercises *Systems of Linear Inequalities*

Graph the system of linear inequalities.

1. $2x + y \leq 4$
$3x - y > 6$

2. $4x - y < 4$
$x + 2y < 2$

3. $x + y > 0$
$x - 3y > 3$

4. $x + 2y > 4$
$2x - y > 6$

5. $2x > y + 3$
$x < 3 - 2y$

6. $2 < 3x - y$
$x - 3y \leq 4$

7. $x \geq 0$
$y \geq 0$
$x \leq 2$
$y \leq 4$

8. $x \geq -1$
$y \geq -1$
$x \leq 1$
$y \leq 2$

9. $x + y \leq 1$
$-x + y \geq 1$
$y \geq 0$

10. $x + 3y < 3$
$x > 0$
$y > 0$

11. $x + y \leq 5$
$x \geq 2$
$y \geq 0$

12. $2x + y \geq 2$
$x \geq 2$
$y \leq 1$

SUMMARIZING KEY IDEAS

Two or more linear inequalities form a system of linear inequalities. You can solve a system of linear inequalities by graphing.

TEXTBOOK LINK

On pages S53–S54, you solved a system of linear inequalities by graphing. In your textbook, Lesson 7.6, you will graph a system of linear inequalities and identify the solution of the system by shading.

Topic Review *Systems of Equations and Inequalities*

These exercises will help you check that you can solve a system of linear equations or a system of linear inequalities. If you have any questions about systems of equations or inequalities, be sure to get them answered before going on to the next section.

Check whether the ordered pair is a solution of the linear system.

1. $x - 3y = -12$
$2x + y = -3$ $\quad (-3, 3)$

2. $4x + y = 5$
$2x + y = -1$ $\quad (-2, 3)$

In Exercises 3–14, use substitution, addition/subtraction, or linear combinations to solve the linear system. Tell which method you chose. Check the solution in each of the original equations.

3. $5g = 3f + 6$
$g - 2f = 46$

4. $-7c + 2d = 31$
$-17c = 17 + 2d$

5. $10r - 9s = 18$
$6s + 2r = 1$

6. $y = -3x + 1$
$15x - 2 = -y$

7. $n = 39 - 3m$
$n = 2m - 6$

8. $2a + 5b = 6$
$3a = 10b + 2$

9. $6m + 12n = 7$
$8m - 15n = -1$

10. $k = 10 - 2j$
$k = 4j + 36$

11. $2x = y + 1$
$5y - 2x = 1$

In Exercises 12 and 13, check whether the ordered pair is a solution of the system of linear inequalities.

12. $5y - 2x > -15$
$3x + 2y \geq 13$ $\quad (5, -1)$

13. $y > x$
$-4x + 12 \leq 2y$ $\quad (3, 4)$

14. Write a system of linear inequalities that defines the shaded region.

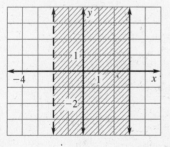

Graph the system of linear inequalities.

15. $3x > 2y - 6$
$x + y > -1$

16. $8x + 6y \geq -24$
$4y - 5x \leq 16$
$y < 2$

17. $y > -x$
$-2x + y < 3$
$y \geq 3$

My Review Notes

Students add, subtract, multiply, and divide monomials and polynomials. Students solve multistep problems, including word problems, by using these techniques.

TEXTBOOK REFERENCES
Developing Concepts 10.1
Lesson 10.1

KEY WORDS
• polynomial
• like terms
• standard form
• opposite

> **STUDY TIP**
> Remember that like terms are terms in an expression that have the same variable raised to the same power.

Add and Subtract Polynomials

ADDING POLYNOMIALS

To add polynomials, combine like terms, then simplify. When adding polynomials, you can use either a vertical format or a horizontal format.

Example 1

Use a vertical format to find the sum
$(3x^3 + 2x^2 - 2x - 5) + (x^2 - 5x + 7)$.
Write the answer in standard form.

▶ **Solution**

$$3x^3 + 2x^2 - 2x - 5 \longleftarrow \text{Line up like terms vertically.}$$
$$+ \quad\quad x^2 - 5x + 7$$
$$\overline{3x^3 + 3x^2 - 7x + 2} \longleftarrow \text{Add like terms in each column.}$$

Example 2

Use a horizontal format to find the sum
$(6x^4 - 2x^3 + 7x^2 + x) + (-7x^4 + 2x^3 - 5x + 7)$.
Write the answer in standard form.

▶ **Solution** Rearrange by grouping like terms.

$(6x^4 - 2x^3 + 7x^2 + x) + (-7x^4 + 2x^3 - 5x + 7)$
$= (6x^4 - 7x^4) + (-2x^3 + 2x^3) + 7x^2 + (x - 5x) + 7$
$= -1x^4 + 0x^3 + 7x^2 + (-4x) + 7$
$= -x^4 + 7x^2 - 4x + 7$

Checkpoint ✓ **Add Polynomials**

Find the sum. Write the answer in standard form.

1. $(a^4 + 6a^2 + 4a^3 + 6 + 7a) + (a^3 + 3a^2 + 5a^4 + 4a + 1)$

2. $(x^2 + 3x^3 + 7x + 8x^4 + 1) + (6x^4 + 3x^3 + 5x + 6 + 3x^2)$

3. $(3d^2 - 6d - 2d^3 + 1) + (-d^3 + 2d - 5d^2 - 5)$

My Notes

TEXTBOOK REFERENCES
Developing Concepts 10.2
Lessons 10.2, 10.3

KEY WORDS
• FOIL pattern
• distributive property
• special product

Multiply Polynomials

USING THE FOIL PATTERN

The FOIL pattern can be used to multiply two binomials: multiply the **F**irst, **O**uter, **I**nner, and **L**ast terms. Then combine like terms.

> **MULTIPLYING TWO BINOMIALS USING THE FOIL PATTERN**
>
> Product of | Product of | Product of
> First terms | Outer terms | Inner terms | Product of
> | | | Last terms
>
> $(x + 5)(x + 3) = x^2 + 3x + 5x + 15$
> $= x^2 + 8x + 15$

Example 1

Find the product $(x + 3)(x + 7)$.

▶ **Solution**

$$(x + 3)(x + 7) = x^2 + 7x + 3x + 21$$
$$= x^2 + 10x + 21 \qquad \text{Combine like terms.}$$

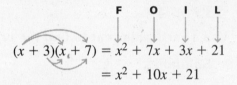

Example 2

Find the product $(3a - b)(a - 2b)$.

▶ **Solution**

$$(3a - b)(a - 2b) = 3a^2 - 6ab - ab + 2b^2$$
$$= 3a^2 - 7ab + 2b^2 \qquad \text{Combine like terms.}$$

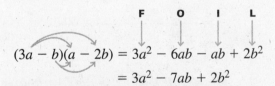

 **Checkpoint ✓** *Use the FOIL Pattern*

Find the product using the FOIL pattern.

  1. $(x + 1)(x - 5)$ **2.** $(y + 2)(y - 4)$ **3.** $(c + d)(e + f)$

My Notes

USING THE DISTRIBUTIVE PROPERTY

The FOIL pattern applies only to the multiplication of two binomials. When multiplying polynomials with more terms, distribute the terms using either a horizontal or vertical format. Each term of one polynomial must be multiplied by each term of the second polynomial.

Example 3

Use a horizontal format to find the product
$(4y - 5)(-2y + y^2 - 3y^3)$.

▶ **Solution**

Multiply $(-2y + y^2 - 3y^3)$ by each term of $(4y - 5)$.

$(4y - 5)(-2y + y^2 - 3y^3)$

$= (4y)(-2y + y^2 - 3y^3) + (-5)(-2y + y^2 - 3y^3)$

$= -8y^2 + 4y^3 - 12y^4 + 10y - 5y^2 + 15y^3$

$= -12y^4 + (15y^3 + 4y^3) + (-8y^2 - 5y^2) + 10y$

$= -12y^4 + 19y^3 - 13y^2 + 10y$

Example 4

Use a vertical format to find the product $(2x - 7)(3 - 4x - x^2)$.

▶ **Solution**

> **STUDY TIP**
> When multiplying vertically, it is helpful to write the polynomials in standard form before multiplying.

Line up like terms vertically. Then multiply as shown below.

$$
\begin{array}{r}
-x^2 - 4x + 3 \qquad \text{Standard form} \\
\times \qquad\qquad 2x - 7 \qquad \text{Standard form} \\
\hline
7x^2 + 28x - 21 \quad \longleftarrow \quad -7(-x^2 - 4x + 3) \\
+ \;\; -2x^3 - 8x^2 + 6x \qquad \longleftarrow \quad 2x(-x^2 - 4x + 3) \\
\hline
-2x^3 - x^2 + 34x - 21 \qquad \text{Combine like terms.}
\end{array}
$$

Checkpoint ✓ *Use the Distributive Property*

Use a horizontal format to find the product.

4. $(5a + 1)(a^2 - 2a + 3)$ **5.** $(8b - 5)(4 - 2b - 7b^2)$

Use a vertical format to find the product.

6. $(2s - 1)(s^2 - 5s + 6)$ **7.** $(11t + 2)(3t^2 + t + 4)$

USING SPECIAL PRODUCTS PATTERNS

A common type of product is the square of a binomial. In its pattern, the middle term is always twice the product of the two terms in the binomial.

SQUARE OF A BINOMIAL PATTERN

$$(a + b)^2 = (a + b)(a + b) = a^2 + ab + ab + b^2 = a^2 + 2ab + b^2$$
$$(a - b)^2 = (a - b)(a - b) = a^2 - ab - ab + b^2 = a^2 - 2ab + b^2$$

Example 5

Find the product.

a. $(2a + 3b)^2$

b. $(3m^2 - 11)^2$

▶ *Solution*

a. $(a + b)^2 = a^2 + 2ab + b^2$ Write pattern.

$(2a + 3b)^2 = (2a)^2 + 2(2a)(3b) + (3b)^2$ Apply pattern.

$= 4a^2 + 12ab + 9b^2$ Simplify.

b. $(a - b)^2 = a^2 - 2ab + b^2$ Write pattern.

$(3m^2 - 11)^2 = (3m^2)^2 - 2(3m^2)(11) + (11)^2$ Apply pattern.

$= 9m^4 - 66m^2 + 121$ Simplify.

Another common type of product is the product of the sum and difference of two terms. This special product has no middle term.

SUM AND DIFFERENCE PATTERN

$$(a + b)(a - b) = a^2 - ab + ab - b^2 = a^2 - b^2$$

SUMMARIZING KEY IDEAS

When you multiply polynomials you can use the FOIL pattern and the distributive property. In special cases you can use the sum and difference pattern or the square of a binomial pattern.

Example 6

Find the product.

a. $(z + 3)(z - 3)$ **b.** $(2r^3 + 5)(2r^3 - 5)$

▶ **Solution**

a. $(a + b)(a - b) = a^2 - b^2$ Write pattern.

$(z + 3)(z - 3) = z^2 - 3^2$ Apply pattern.

$= z^2 - 9$ Simplify.

b. $(a + b)(a - b) = a^2 - b^2$ Write pattern.

$(2r^3 + 5)(2r^3 - 5) = (2r^3)^2 - (5)^2$ Apply pattern.

$= 4r^6 - 25$ Simplify.

Checkpoint ✓ Use Special Products Patterns

Find the product using a special product pattern.

8. $(m - n)^2$ **9.** $(u + v)(u - v)$ **10.** $(3p + 4)^2$

11. $(7m + 1)(7m - 1)$ **12.** $(2t - 10)^2$ **13.** $(x - 11)(x + 11)$

Exercises Multiply Polynomials

TEXTBOOK LINK

On pages S59–S62, you multiplied polynomials by using the FOIL pattern and by using the distributive property. You also found the square of a binomial and found the product of the sum and difference of two terms. In your textbook, Lessons 10.2 and 10.3, you will use the distributive property and the FOIL pattern to multiply binomials in both vertical and horizontal formats. Then you will use special product patterns to multiply polynomials.

Find the product.

1. $(-2y + y^2 - 4y^3)(6y - 3)$ **2.** $(-5x + x^2 - 3x^3)(4x - 6)$

3. $(4z + z^2 - 5z^3)(2z + 3)$ **4.** $(8m + 3m^2 - 2m^3)(m + 5)$

5. $(x^2 + 3x - 5)(x + 2)$ **6.** $(y^2 + 11y - 12)(y + 5)$

7. $(z - 3)(z^2 - 5z - 12)$ **8.** $(w - 4)(w^2 - 5w + 1)$

9. $(c^2 + 2c + 1)(c - 5)$ **10.** $(3 + k)(7k - k^2 + 8)$

11. $(2t - u)^2$ **12.** $(h + 9)(h - 9)$ **13.** $(3j - h)^2$

14. $(x + 4)(x - 4)$ **15.** $(d^2 + e^2)^2$ **16.** $(y + 5)(y - 5)$

17. $(mn + k)(mn - k)$ **18.** $(2 - 6x^3y^2)^2$ **19.** $(q^2 + p^2)(q^2 - p^2)$

20. $(1 - 12g^3h^2)^2$ **21.** $(4s + 5t)(4s - 5t)$ **22.** $(5ab^3 + 6c^2d^4)^2$

TEXTBOOK REFERENCES

Lesson 11.3

KEY WORDS
- standard form
- monomial
- binomial

Divide Polynomials

To divide a polynomial by a monomial, divide each term of the polynomial by the monomial and simplify. When dividing polynomials, the polynomials need to be written in standard form.

Example 1

Divide $(9w^4 + 3w^3 - w^2)$ **by** $3w^3$.

▶ **Solution**

$$\frac{9w^4 + 3w^3 - w^2}{3w^3} = \frac{9w^4}{3w^3} + \frac{3w^3}{3w^3} - \frac{w^2}{3w^3}$$ Divide each term by $3w^3$.

$$= \frac{3 \cdot 3 \cdot w^3 \cdot w}{3 \cdot w^3} + \frac{3w^3}{3w^3} - \frac{w^2}{3 \cdot w^2 \cdot w}$$ Find common factors.

$$= \frac{\cancel{3} \cdot 3 \cdot \cancel{w^3} \cdot w}{\cancel{3} \cdot \cancel{w^3}} + \frac{\cancel{3w^3}}{\cancel{3w^3}} - \frac{\cancel{w^2}}{3 \cdot \cancel{w^2} \cdot w}$$ Divide out common factors.

$$= 3w + 1 - \frac{1}{3w}$$ Simplify the expression.

Check ✓ Check the result of a division by using multiplication.

$$\left(3w + 1 - \frac{1}{3w}\right)(3w^3) = 9w^4 + 3w^3 - w^2 ✓$$

When you divide one polynomial by another polynomial, factor if possible and then divide by any common factors.

Example 2

Divide $(a^2 + 3a - 4)$ **by** $(a + 4)$.

▶ **Solution**

$$\frac{a^2 + 3a - 4}{(a + 4)} = \frac{(a - 1)(a + 4)}{(a + 4)}$$ Factor the numerator.

$$= \frac{(a - 1)\cancel{(a + 4)}}{\cancel{(a + 4)}}$$ Divide out the common factor $(a + 4)$.

$$= a - 1$$ Simplify the expression.

My Notes

Checkpoint ✓ *Divide Polynomials*

Find the quotient.

1. Divide $(15k^6 - 5k^5 + 60k^4)$ by $5k^2$.

2. Divide $(3m^4 + 4m^3 - 7m^2)$ by m.

3. Divide $(p^2 + 5p - 36)$ by $(p + 9)$.

4. Divide $(2n^2 - n - 3)$ by $(2n - 3)$.

SUMMARIZING KEY IDEAS

To divide a polynomial by a monomial, divide each term of the polynomial by the monomial. When you divide one polynomial by another, factor if possible and then divide by any common factors.

Exercises *Divide Polynomials*

Find the quotient.

1. Divide $(18s^2 - 51s)$ by $3s$. **2.** Divide $(108a^3 - 72a^4)$ by $9a^3$.

3. Divide $(6x^4 + 3x^3 - x^2)$ by $3x^3$. **4.** Divide $(10z^4 + 5z^3 - z^2)$ by $5z$.

5. Divide $(x^3 - 18x^2 + 3x - 7)$ by x.

6. Divide $(3y^4 - 15y^3 + 21y^2 + 6y - 9)$ by $3y$.

7. Divide $(4m^4 + 6m^3 - 20m^2 + 8m - 12)$ by $2m$.

8. Divide $(18p^5q^4 - 54p^3q^3 + 27p^2q^5 - 9pq^2)$ by $9pq^2$.

9. Divide $(21a^3b^2 - 56a^5b - 28a^2b + 63a^5b^2)$ by $7a^2b$.

Find the quotient.

10. Divide $(x^2 - 6x + 5)$ by $(x - 1)$.

11. Divide $(n^2 + n - 30)$ by $(n + 6)$.

12. Divide $(3j^2 - 14j - 5)$ by $(j - 5)$.

13. Divide $(3p^2 + 4p - 4)$ by $(p + 2)$.

14. Divide $(k^2 + 9k + 18)$ by $(k + 6)$.

15. Divide $(3t^2 - 10t - 8)$ by $(3t + 2)$.

16. Divide $(10h^2 - 9h - 7)$ by $(2h + 1)$.

17. Divide $(15x^2 + 7xy - 2y^2)$ by $(5x - y)$.

18. Divide $(10m^2 - 11mn + 3n^2)$ by $(2m - n)$.

TEXTBOOK LINK

On pages S63 and S64, you divided polynomials. To divide, you factored if it was possible and then divided by the common factors. In your textbook, Lesson 11.3, you will simplify quotients of polynomials by dividing out the common factors.

Topic Review Polynomials

These exercises will help you check that you can add, subtract, multiply, and divide polynomials. If you have any questions about performing operations on polynomials, be sure to get them answered before going on to the next section.

Use a vertical format to find the sum, difference, or product. Write the answer in standard form.

1. $(4x^2 + 9x + 11) - (x^2 - 3)$ **2.** $(3x^3 + 2x^2 - 9)(5x - 8)$

3. $(9x^3 + 2x^2 + 7x) - (x^2 + x)$ **4.** $(5x^2 - 6x - 2)(4x^2 + 7)$

5. $(5x^4 + 4x^3 + 7x) + (x^4 + 3x)$ **6.** $(x^4 - 7x^3 - x^2) - (x^3 + 4x^2)$

Use a horizontal format to find the sum, difference, or product. Write the answer in standard form.

7. $(6s^2 - 5s + 1)(3s^2 + 2)$ **8.** $(3x^2 - 5) + (7x^2 - 4x - 1)$

9. $(5f^2 + 7f - 13) + (f^2 + 2)$ **10.** $(11m^2 + 2m) - (m^2 - m - 4)$

11. $(4d^3 + 7d^2 - 10d)(2d^2 - 1)$ **12.** $(n^3 + 6n^2 - 2n) - (n^3 - 8n)$

Find the product.

13. $(2a + 5)(2a - 5)$ **14.** $(6x + 5)^2$ **15.** $(h + 1)(3h + 7)$

16. $(4g - 1)(5g - 3)$ **17.** $(7p - 8)^2$ **18.** $(9s + 1)(9s - 1)$

Find the quotient.

19. Divide $(12w^4 - 30w^2 + 42w)$ by $6w$.

20. Divide $(30q^6 + 4q^5 - 2q^4 - 10q^3)$ by $2q^4$.

21. Divide $(2t^2 - 7t - 49)$ by $(2t + 7)$.

22. Divide $(n^4 + 11n^2 + 30)$ by $(n^2 + 5)$.

23. Find the perimeter of a rectangle with length $(7x + 12)$ and width $(x^2 - 3x - 5)$.

24. You are tiling a kitchen floor that has length $5x$ and width $(2x + 11)$.

   a. Write an expression for the area of the floor.

   b. Each tile has area $0.2x$. Write an expression that gives the number of tiles needed to cover the entire kitchen floor.

My Review Questions

KEY STANDARD

Algebra 1 **13.0** Add, subtract, multiply, and divide rational expressions and functions.

TEXTBOOK REFERENCES
Lesson 11.4

KEY WORDS
• rational expression
• reciprocal
• divisor
• order of operations

Multiplying and Dividing Rational Expressions

MULTIPLYING RATIONAL EXPRESSIONS

Recall that a rational expression is an expression that can be written in the form $\dfrac{p}{q}$ where p and q are polynomials and $q \neq 0$. The rule for multiplying rational expressions is similar to the rule for multiplying rational numbers.

> **STUDY TIP**
> In Example 1, you do not need to write the prime factorizations of 60 and 6 if you recognize 6 as their greatest common factor.

MULTIPLYING RATIONAL EXPRESSIONS

If $\dfrac{p}{q}$ and $\dfrac{r}{s}$ are rational expressions, then $\dfrac{p}{q} \cdot \dfrac{r}{s} = \dfrac{p \cdot r}{q \cdot s}$.

Example 1

Simplify $\dfrac{4b}{3a^3} \cdot \dfrac{15ab^2}{2b}$.

▶ **Solution**

$$\frac{4b}{3a^3} \cdot \frac{15ab^2}{2b} = \frac{60ab^3}{6a^3b} = \frac{\cancel{6} \cdot 10 \cdot \cancel{a} \cdot b \cdot b \cdot \cancel{b}}{\cancel{6} \cdot \cancel{a} \cdot a \cdot a \cdot \cancel{b}} = \frac{10b^2}{a^2}$$

Example 2

Simplify $\dfrac{2x^2 - 5x - 12}{6x} \cdot \dfrac{-3x - 12}{x^2 - 16}$.

▶ **Solution**

$$\frac{2x^2 - 5x - 12}{6x} \cdot \frac{-3x - 12}{x^2 - 16} = \frac{(2x + 3)(x - 4)}{2 \cdot 3 \cdot x} \cdot \frac{-3(x + 4)}{(x - 4)(x + 4)}$$

$$= \frac{-3(2x + 3)(x - 4)(x + 4)}{2 \cdot 3 \cdot x(x - 4)(x + 4)}$$

$$= \frac{-(2x + 3)}{2x} = \frac{-2x - 3}{2x}$$

My Notes

Checkpoint ✓ *Multiplying Rational Expressions*

Write the product in simplest form.

1. $\dfrac{b^3}{4} \cdot \dfrac{42a}{12ab^2}$

2. $\dfrac{r^2 + 3r - 10}{2r + 1} \cdot \dfrac{2r^2 + 7r + 3}{r + 5}$

DIVIDING RATIONAL EXPRESSIONS

Dividing rational expressions is similar to dividing fractions.

> ### DIVIDING RATIONAL EXPRESSIONS
>
> If $\dfrac{p}{q}$ and $\dfrac{r}{s}$ are rational expressions and $q \neq 0$, $r \neq 0$ and
>
> $s \neq 0$, then $\dfrac{p}{q} \div \dfrac{r}{s} = \dfrac{p}{q} \cdot \dfrac{s}{r} = \dfrac{p \cdot s}{q \cdot r}$.

> **STUDY TIP**
> Multiplication and division are inverse operations. Reciprocals are two numbers whose product is 1. For example, $\dfrac{3}{4} \cdot \dfrac{4}{3} = 1$.

Example 3

Simplify $\dfrac{3}{a} \div \dfrac{6b}{5a}$.

▶ **Solution**

$$\dfrac{3}{a} \div \dfrac{6b}{5a} = \dfrac{3}{a} \cdot \dfrac{5a}{6b} = \dfrac{15a}{6ab} = \dfrac{\cancel{3} \cdot 5 \cdot \cancel{a}}{2 \cdot \cancel{3} \cdot \cancel{a} \cdot b} = \dfrac{5}{2b}$$

Example 4

Simplify $\dfrac{x^2 + x - 2}{3x^2 + 9x + 6} \div (x - 1)$.

▶ **Solution**

$$\dfrac{x^2 + x - 2}{3x^2 + 9x + 6} \div (x - 1) = \dfrac{x^2 + x - 2}{3x^2 + 9x + 6} \cdot \dfrac{1}{x - 1}$$

$$= \dfrac{(x + 2)(x - 1)}{3(x + 2)(x + 1)} \cdot \dfrac{1}{x - 1}$$

$$= \dfrac{(x + 2)(x - 1)}{3(x + 2)(x + 1)(x - 1)} = \dfrac{1}{3(x + 1)}$$

My Notes

Dividing Rational Expressions

Write the quotient in simplest form.

3. $\dfrac{2b}{5a^2} \div \dfrac{6b^2}{10a}$

4. $\dfrac{r^2 + 9r + 20}{r^2 - 64} \div \dfrac{r + 5}{r + 8}$

COMBINING MULTIPLICATION AND DIVISION

Rational expressions involving both multiplication and division should be simplified using order of operations.

Example 5

Simplify $\dfrac{2m^2 - 5m - 3}{9 - m^3} \div \dfrac{4m + 2}{2m^2 + 2m - 12} \cdot \dfrac{2}{m - 2}$.

▶ *Solution*

$\dfrac{2m^2 - 5m - 3}{9 - m^2} \div \dfrac{4m + 2}{2m^2 + 2m - 12} \cdot \dfrac{2}{m - 2}$

$= \dfrac{2m^2 - 5m - 3}{9 - m^2} \cdot \dfrac{2m^2 + 2m - 12}{4m + 2} \cdot \dfrac{2}{m - 2}$

$= \dfrac{(2m + 1)(m - 3)}{-1(m - 3)(m + 3)} \cdot \dfrac{2(m + 3)(m - 2)}{2(2m + 1)} \cdot \dfrac{2}{m - 2}$

$= \dfrac{2 \cdot 2(2m + 1)(m - 3)(m + 3)(m - 2)}{-2(m - 3)(m + 3)(2m + 1)(m - 2)}$

$= \dfrac{2 \cdot 2\cancel{(2m + 1)}\cancel{(m - 3)}\cancel{(m + 3)}\cancel{(m - 2)}}{-2\cancel{(m - 3)}\cancel{(m + 3)}\cancel{(2m + 1)}\cancel{(m - 2)}}$

$= \dfrac{2}{-1} = -2$

Combining Multiplication and Division

Write the expression in simplest form.

5. $\dfrac{x^2 - 9}{2x - 6} \div \dfrac{x^2 + 8x + 15}{3x^2 - 6x - 45} \cdot \dfrac{2x^2 + 9x - 5}{x^2 - 2x - 15}$

Exercises Multiply and Divide Rational Expressions

Write the product in simplest form.

1. $\dfrac{15p^3}{pq} \cdot \dfrac{2q}{75pq}$

2. $\dfrac{2a^2b}{7b^3} \cdot \dfrac{35b}{a^3}$

3. $\dfrac{m-2}{3m+9} \cdot \dfrac{2m+6}{2m-4}$

4. $\dfrac{x-5}{4x+6} \cdot \dfrac{6x+9}{3x-15}$

5. $\dfrac{2x^2-2x-4}{8x} \cdot \dfrac{-8x-16}{x^2-4}$

6. $\dfrac{r^2-6r+8}{2r} \cdot \dfrac{-2r-8}{r^2-16}$

7. $\dfrac{9a^2+43a-10}{3a+2} \cdot \dfrac{12a+8}{a+5}$

8. $\dfrac{10l^2-67l-21}{20l^2-84l-27} \cdot \dfrac{6l-27}{l^2-49}$

Write the quotient in simplest form.

9. $\dfrac{18a^3b^2}{25b^2} \div \dfrac{12a^2c}{5b}$

10. $\dfrac{24x^5y^3}{18z^2} \div \dfrac{15x^2y}{12z}$

11. $\dfrac{4h^2-25}{9h^2-4} \div \dfrac{4h^2-16h+15}{6h^2-5h-6}$

12. $\dfrac{8a+12}{3a+3} \div \dfrac{2a^2+5a-3}{2a^2+a-1}$

13. $\dfrac{x^2-9}{x^2-6x+5} \div \dfrac{2x+6}{x^2-25}$

14. $\dfrac{2a^2-ab-6b^2}{2b^2+9ab-5a^2} \div \dfrac{a+2b}{a^2-4b^2}$

Write the expression in simplest form.

15. $\dfrac{12s^2+19s+5}{2s^2-7s+3} \div \dfrac{10s+5}{3s-2} \cdot \dfrac{2s^2-5s-3}{12s^2-5s-3}$

16. $\dfrac{14y^2+13y+3}{30y^2-27y-21} \div \dfrac{6y^2+11y-10}{25y^2-50y+21} \cdot \dfrac{2y+5}{5y}$

17. $\dfrac{m^2-9n^2}{4m^2-6mn} \div \dfrac{m^2-n^2}{4m^2-9n^2} \cdot \dfrac{2m^2+9mn+9n^2}{m^2+3mn-18n^2}$

Find the volume of the rectangular solid with the given dimensions. ($V = lwh$)

18. length: $\dfrac{x-2}{x^2+2x-35}$; width: $\dfrac{3x+2}{4}$; height: $\dfrac{x-5}{3x+2}$

19. length: $\dfrac{3x^2+8x-3}{2x^2+5x-3}$; width $\dfrac{4x-6}{3x-1}$; height: $\dfrac{2x-1}{2x^2+7x-15}$

TEXTBOOK LINK

On pages S66–S69, you multiplied and divided rational expressions. In your textbook, Lesson 11.4, you will multiply and divide rational expressions.

TEXTBOOK REFERENCES
Lessons 11.5 and 11.6

KEY WORDS
- rational expression
- common denominator
- least common denominator (LCD)

My Notes

Adding and Subtracting Rational Expressions

EXPRESSIONS WITH LIKE DENOMINATORS

Rational expressions are added or subtracted in the same way as fractions. As with fractions, you may need to find the LCD of the rational expressions before you add or subtract. The method for adding or subtracting rational expressions may be generalized as follows.

ADDITION AND SUBTRACTION OF RATIONAL EXPRESSIONS

If $\dfrac{p}{q}$ and $\dfrac{r}{q}$ are rational expressions with $q \neq 0$, then

$$\frac{p}{q} + \frac{r}{q} = \frac{p+r}{q} \text{ and } \frac{p}{q} - \frac{r}{q} = \frac{p-r}{q}.$$

Example 1

Add $\dfrac{4}{3m} + \dfrac{2}{3m}$.

▶ *Solution*

$$\frac{4}{3m} + \frac{2}{3m} = \frac{4+2}{3m} = \frac{6}{3m} = \frac{2}{m}$$

Example 2

Subtract $\dfrac{5n}{n+2} - \dfrac{n-8}{n+2}$.

▶ *Solution*

$$\frac{5n}{n+2} - \frac{n-8}{n+2} = \frac{5n-(n-8)}{n+2}$$
Subtract the numerators.

$$= \frac{5n-n+8}{n+2}$$
Distribute the negative.

$$= \frac{4n+8}{n+2}$$
Combine like terms.

$$= \frac{4(n+2)}{n+2} = 4$$
Factor numerator and simplify.

▶**STUDY TIP**
When subtracting, it is often helpful to use parentheses when you rewrite polynomial numerators to clearly indicate the expressions being subtracted.

Checkpoint ✓ **Add and Subtract with Like Denominators**

Write the sum or difference in simplest form.

1. $\dfrac{m}{m-3} + \dfrac{2}{m-3}$ **2.** $\dfrac{3c}{5c} - \dfrac{c-c^2}{5c}$ **3.** $\dfrac{6r}{2rs} + \dfrac{4r}{2rs}$

EXPRESSIONS WITH UNLIKE DENOMINATORS

To add or subtract rational expressions with unlike denominators, first write them as equivalent expressions with a least common denominator (LCD).

Example 3

Simplify $\dfrac{3}{4a} + \dfrac{5}{6a^2} - \dfrac{1}{a}$.

▶ **Solution**

$$\dfrac{3}{4a} + \dfrac{5}{6a^2} - \dfrac{1}{a} = \dfrac{3}{4a} \cdot \dfrac{3a}{3a} + \dfrac{5}{6a^2} \cdot \dfrac{2}{2} - \dfrac{1}{a} \cdot \dfrac{12a}{12a} \qquad \text{Rewrite using LCD.}$$

$$= \dfrac{9a}{12a^2} + \dfrac{10}{12a^2} - \dfrac{12a}{12a^2} \qquad \text{Simplify.}$$

$$= \dfrac{9a + 10 - 12a}{12a^2} = \dfrac{10 - 3a}{12a^2} \qquad \text{Subtract like terms.}$$

Example 4

Simplify $\dfrac{2y+1}{9-y^2} + \dfrac{2}{y-3} - \dfrac{1}{y+3}$.

▶ **Solution**

$$\dfrac{2y+1}{9-y^2} + \dfrac{2}{y-3} - \dfrac{1}{y+3}$$

$$= \dfrac{-(2y+1)}{(y-3)(y+3)} + \dfrac{2(y+3)}{(y-3)(y+3)} - \dfrac{1(y-3)}{(y+3)(y-3)}$$

$$= \dfrac{-(2y+1)}{(y-3)(y+3)} + \dfrac{2y+6}{(y-3)(y+3)} - \dfrac{y-3}{(y+3)(y-3)}$$

$$= \dfrac{-2y - 1 + 2y + 6 - (y-3)}{(y-3)(y+3)}$$

$$= \dfrac{8-y}{(y-3)(y+3)}$$

My Notes

Checkpoint ✓ **Add and Subtract with Unlike Denominators**

Simplify the expression.

4. $\dfrac{g}{g+1} + \dfrac{g+2}{g} - \dfrac{2}{g^2+g}$

5. $\dfrac{2e}{2e+1} - \dfrac{1}{e} + \dfrac{e}{2e^2+e}$

SUMMARIZING KEY IDEAS

Rational expressions are added or subtracted in the same way as fractions. To add or subtract rational expressions with unlike denominators, first write them as equivalent expressions with a least common denominator (LCD).

Exercises **Add and Subtract Rational Expressions**

Write the sum or difference in simplest form.

1. $\dfrac{a}{a+7} + \dfrac{1}{a+7}$

2. $\dfrac{2d}{8d^2} - \dfrac{18d}{8d^2}$

3. $\dfrac{3b}{12ab} + \dfrac{15b}{12ab}$

4. $\dfrac{3}{b-3} - \dfrac{b}{b-3}$

5. $\dfrac{6-x}{4-x^2} - \dfrac{4}{4-x^2}$

6. $\dfrac{8x}{x^3-x} - \dfrac{6x}{x^3-x}$

Simplify the expression.

7. $\dfrac{n+5}{4n+16} + \dfrac{n-1}{3n-9}$

8. $\dfrac{n^2+1}{n^2-2n-15} + \dfrac{n-1}{n+3}$

9. $\dfrac{5x+1}{25-x^2} - \dfrac{5}{x-5}$

10. $\dfrac{a}{6a^2-7a-20} - \dfrac{2}{3a^2-2a-8}$

11. $\dfrac{r+3}{4r} + \dfrac{r+2}{3r^2} + \dfrac{r-4}{12r^2}$

12. $\dfrac{7d-2}{d^2+2d-8} - \dfrac{4}{d+4} - \dfrac{d}{d-2}$

13. $\dfrac{b-2}{2b} + \dfrac{b+3}{3b} - \dfrac{b-2}{6b^2}$

14. $\dfrac{i-24}{i^2-3i-18} - \dfrac{3}{i+3} + \dfrac{i}{i-6}$

15. **ERROR ANALYSIS** Find and correct the error.

$$\frac{2}{c+5} - \frac{3}{c-5} + \frac{1}{(c+5)(c-5)}$$

$$= \frac{2(c-5)}{(c+5)(c-5)} - \frac{3(c+5)}{(c+5)(c-5)} + \frac{1}{(c+5)(c-5)}$$

$$= \frac{2c-10}{(c+5)(c-5)} - \frac{3c+15}{(c+5)(c-5)} + \frac{1}{(c+5)(c-5)}$$

$$= \frac{2c-10-3c+15+1}{(c+5)(c-5)} = \frac{-c+6}{(c+5)(c-5)}$$

TEXTBOOK LINK

📖 On pages S70–S72 you added and subtracted rational expressions. In your textbook, Lessons 11.5 and 11.6, you will add and subtract rational expressions with like denominators and unlike denominators.

Topic Review Rational Expressions

These exercises will help you check that you can add, subtract, multiply, and divide rational expressions and functions. If you have any questions about performing operations on rational expressions, be sure to get them answered before going on to the next section.

Write the expression in simplified form.

1. $\dfrac{2c}{3c^2} + \dfrac{4c}{3c^2}$

2. $\dfrac{4r + 1}{9 - r^2} \div \dfrac{4}{r - 3}$

3. $\dfrac{u - 5}{2u + 4} \div \dfrac{u + 3}{3u + 6} \cdot \dfrac{6}{3u - 15}$

4. $\dfrac{u - 5}{2u + 4} + \dfrac{u + 9}{2u + 4}$

5. $\dfrac{7t^2 - 28t}{2t^2 - 5t - 12} \cdot \dfrac{6t^2 - t - 15}{49t^3}$

6. $\dfrac{2a^2}{6a} - \dfrac{8a}{6a}$

7. $\dfrac{3a - 2}{a^2 - a - 12} \cdot \dfrac{a + 3}{a + 4} \div \dfrac{5}{a^2 - 16}$

8. $\dfrac{3r}{r^2 - 9} - \dfrac{5r}{r - 3}$

9. $\dfrac{18x^2 - 6x}{3x^4} \cdot \dfrac{10x^2 - 7x - 12}{15x^2 + 7x - 4}$

10. $\dfrac{a}{a + 3} - \dfrac{3}{a + 5}$

11. $\dfrac{n - 2}{n^2 + n - 6} + \dfrac{n + 5}{3n^2 + 13n + 12}$

12. $\dfrac{t}{t^2 - 25} - \dfrac{-5}{t^2 - 25}$

13. $\dfrac{n - 2}{n^2 + n - 6} \div \dfrac{n + 5}{3n^2 + 13n + 12}$

14. $\dfrac{5c}{c + 7} - \dfrac{c - 28}{c + 4}$

My Review Questions

In Exercises 15–18, use the rectangle below.

15. Find an expression for the area of the rectangle.

16. Find an expression for the perimeter of the rectangle.

17. What is the area of the rectangle when $x = 2$?

18. What is the perimeter of the rectangle when $x = 2$?

$\dfrac{4}{3x}$

$\dfrac{5}{3x}$

Algebra 1 **14.0** Students solve a quadratic equation by factoring or completing the square.

TEXTBOOK REFERENCES
Developing Concepts 12.5
Lessons 9.2, 9.3, 12.5

KEY WORDS
• perfect square
• square root
• quadratic equation
• completing the square

Quadratic Equations

SOLVE BY FINDING SQUARE ROOTS

When an equation contains a perfect square on one side and a nonnegative constant on the other, you can solve the equation by finding square roots.

FINDING SQUARE ROOTS

For any real number d, $d > 0$, $x^2 = d$ has two solutions, $\pm\sqrt{d}$.

Example 1

Solve $4x^2 = 20$.

▶ **Solution**

$4x^2 = 20$	Write original equation.
$x^2 = 5$	Divide each side by 4.
$x = \pm\sqrt{5}$	Find square roots.

▶ The solutions are $\sqrt{5}$ and $-\sqrt{5}$.

Check ✓ $\quad 4(\sqrt{5})^2 \stackrel{?}{=} 20 \qquad 4(-\sqrt{5})^2 \stackrel{?}{=} 20$

$\qquad\qquad\qquad 4(5) \stackrel{?}{=} 20 \qquad\qquad 4(5) \stackrel{?}{=} 20$

$\qquad\qquad\qquad\quad 20 = 20 ✓ \qquad\qquad\quad 20 = 20 ✓$

Example 2

Solve $(y - 7)^2 = 64$.

▶ **Solution**

$(y - 7)^2 = 64$	Write original equation.
$y - 7 = \pm 8$	Find square root of each side.
$y = 7 \pm 8$	Add 7 to each side.

▶ The solutions are 15 and -1.

Check ✓ $\quad (15 - 7)^2 \stackrel{?}{=} 64 \qquad [(-1) - 7]^2 \stackrel{?}{=} 64$

$\qquad\qquad\qquad\qquad 64 = 64 ✓ \qquad\qquad\qquad 64 = 64 ✓$

My Notes

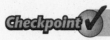 *Solving Quadratic Equations*

Solve the equation.

1. $p^2 = \dfrac{4}{25}$ **2.** $n^2 = \dfrac{1}{100}$ **3.** $2n^2 - 13 = -1$

SOLVE BY COMPLETING THE SQUARE

Completing the square is one method of solving quadratic equations that works whether or not the equation can be factored. That is, it works for all quadratic equations.

COMPLETING THE SQUARE

To complete the square of the expression $x^2 + bx$, add the square of half the coefficient of x, that is, add $\left(\dfrac{b}{2}\right)^2$.

$$x^2 + bx + \left(\dfrac{b}{2}\right)^2 = \left(x + \dfrac{b}{2}\right)^2$$

Example 3

Solve $x^2 + 6x = 40$ by completing the square.

▶ **Solution**

$x^2 + 6x = 40$	Write original equation.
$x^2 + 6x + 3^2 = 40 + 3^2$	Add $\left(\dfrac{6}{2}\right)^2$, or 3^2, to each side.
$(x + 3)^2 = 49$	Write left side as perfect square.
$x + 3 = \pm 7$	Find square root of each side.
$x = -3 \pm 7$	Subtract 3 from each side.
$x = 4$ or $x = -10$	Simplify.

▶ The solutions are 4 and -10. Check these in the original equation to confirm that both are solutions.

▶ **STUDY TIP**
When completing the square to solve an equation, remember that you must always add the term $\left(\dfrac{b}{2}\right)^2$ to both sides of the equation.

Example 4

Solve $y^2 - y - \dfrac{3}{4} = 0$ by completing the square.

▶ *Solution*

$$y^2 - y = \frac{3}{4}$$ Add $\dfrac{3}{4}$ to each side.

$$y^2 - y + \left(\frac{1}{2}\right)^2 = \frac{3}{4} + \left(\frac{1}{2}\right)^2$$ Add $\left(\dfrac{-1}{2}\right)^2$, or $\left(\dfrac{1}{2}\right)^2$, to each side.

$$\left(y - \frac{1}{2}\right)^2 = 1$$ Write left side as perfect square.

$$y - \frac{1}{2} = \pm 1$$ Find square root of each side.

$$y = \frac{1}{2} \pm 1$$ Add $\dfrac{1}{2}$ to each side.

▶ The solutions are $\dfrac{3}{2}$ and $-\dfrac{1}{2}$. Check these in the original equation.

Example 5

Solve $2t^2 + 3t - 4 = 0$ by completing the square.

▶ *Solution*

$$2t^2 + 3t = 4$$ Add 4 to each side.

$$t^2 + \frac{3}{2}t = 2$$ Divide each term by 2.

$$t^2 + \frac{3}{2}t + \left(\frac{3}{4}\right)^2 = 2 + \left(\frac{3}{4}\right)^2$$ Add $\left(\dfrac{3}{2} \div 2\right)^2$, or $\left(\dfrac{3}{4}\right)^2$, to each side.

$$\left(t + \frac{3}{4}\right)^2 = \frac{41}{16}$$ Write left side as perfect square.

$$t + \frac{3}{4} = \pm \frac{\sqrt{41}}{4}$$ Find square root of each side.

$$t = -\frac{3}{4} \pm \frac{\sqrt{41}}{4} = \frac{-3 \pm \sqrt{41}}{4}$$ Subtract $\dfrac{3}{4}$ from each side and simplify.

▶ The solutions are $\dfrac{-3 + \sqrt{41}}{4}$ and $\dfrac{-3 - \sqrt{41}}{4}$. Check these in the original equation to confirm that both are solutions.

> ▶ **STUDY TIP**
> In Example 5, divide each side by 2 so that the coefficient of the t^2-term is 1. Then use the method for completing the square.

Checkpoint ✓ **Solve by Completing the Square**

Solve the equation by completing the square.

4. $x^2 - 8x = -15$ **5.** $x^2 + 18x + 56 = 0$ **6.** $4n^2 = 4n + 1$

Exercises   Solve Quadratic Equations

Solve the equation.

1. $(n - 9)^2 = 27$ **2.** $\left(m + \dfrac{2}{3}\right)^2 = \dfrac{1}{9}$ **3.** $\left(n + \dfrac{3}{4}\right)^2 = \dfrac{1}{16}$

4. $\left(r + \dfrac{2}{3}\right)^2 = \dfrac{5}{9}$ **5.** $\left(y + \dfrac{3}{4}\right)^2 = \dfrac{3}{16}$ **6.** $2(5x + 1)^2 = 50$

7. $3(2s + 4)^2 = 36$ **8.** $5(z - 3)^2 = 35$ **9.** $7(m + 4)^2 = 70$

Solve the equation by completing the square.

10. $x^2 - 4x + 2 = 0$ **11.** $n^2 - 3n + \dfrac{5}{4} = 0$ **12.** $r^2 - 3r - 10 = 0$

13. $p^2 + 5p + 6 = 0$ **14.** $4y^2 - 6y - \dfrac{1}{2} = 0$ **15.** $3b^2 - 12b - 9 = 0$

16. $3n^2 - 8n + 4 = 0$ **17.** $2t^2 - 5t = 4$ **18.** $2c^2 + c = 5$

Use a calculator to solve the equation by completing the square. Round your answers to the nearest hundredth.

19. $t^2 - 3.7t - 10 = 0$ **20.** $p^2 - 5.2p + 6 = 0$

21. $y^2 = 10 + 5.4y$ **22.** $b^2 = 2\sqrt{2}\,b + 2$

23. ERROR ANALYSIS Find and correct the error.

$$x^2 + 14x - 3 = 0$$
$$x^2 + 14x = 3$$
$$x^2 + 14x + 7^2 = 3$$
$$(x + 7)^2 = 3$$
$$x + 7 = \pm\sqrt{3}$$
$$x = -7 \pm \sqrt{3}$$

Students apply algebraic techniques to solve rate problems, work problems, and percent mixture problems.

TEXTBOOK REFERENCES

Lessons 3.8, 11.6, 11.7

KEY WORDS

• rate

Rate Problems

To solve rate problems involving distance and time, you can use the distance formula and solve for the missing variable.

$$\boxed{\text{distance}} = \boxed{\text{rate}} \cdot \boxed{\text{time}} \quad \text{or} \quad \boxed{\text{time}} = \frac{\boxed{\text{distance}}}{\boxed{\text{rate}}}$$

Example 1

A sled dog racer stopped while crossing some difficult terrain to tend to one of his dogs. He traveled 12 miles before stopping, and then traveled 48 miles after stopping, at twice the earlier rate. If the actual running time was 6 hours, find his average rates before and after stopping.

▶ **Solution**

Use a table to organize the data. Let r = average rate before stopping. Then $2r$ = average rate after stopping.

	Distance (mi)	Rate (mi/h)	Time (h)
Before stopping	12	r	$\dfrac{12}{r}$
After stopping	48	$2r$	$\dfrac{48}{2r}$

My Notes

$$\frac{12}{r} + \frac{48}{2r} = 6 \qquad \text{The sum of the two travel times is 6 hours.}$$

$$2r\left(\frac{12}{r} + \frac{48}{2r}\right) = 2r(6) \qquad \text{Multiply by the LCD, 2r.}$$

$$2r\left(\frac{12}{\cancel{r}}\right) + \cancel{2r}\left(\frac{48}{\cancel{2r}}\right) = 2r(6) \qquad \text{Use distributive property.}$$

$$24 + 48 = 12r \qquad \text{Simplify.}$$

$$72 = 12r \qquad \text{Simplify.}$$

$$6 = r \qquad \text{Divide each side by 12.}$$

▶ The average rate before stopping was 6 miles per hour.
The average rate after stopping was $2r = 2(6) = 12$ miles per hour.

▶STUDY TIP
When you travel *downstream*, you are going with the current. When you travel *upstream*, you are going against the current and your rate of speed will be less.

Example 2

The speed of a boat in still water is 30 miles per hour. It travels 80 miles downstream in the same time that it takes to travel 40 miles upstream. Find the rate of the stream's current.

▶ Solution

Use a table to organize the data. Let c = rate of stream. Then $30 + c$ = rate of boat downstream and $30 - c$ = rate of boat upstream.

	Distance (mi)	Rate (mi/h)	Time (h)
Boat travels downstream	80	$30 + c$	$\dfrac{80}{30 + c}$
Boat travels upstream	40	$30 - c$	$\dfrac{40}{30 - c}$

Multiply the denominators of each fraction to get the least common denominator (LCD): $(30 - c)(30 + c)$.

Travel time for 40 miles upstream equals travel time for 80 miles downstream.

$$\frac{40}{30 - c} = \frac{80}{30 + c}$$ Write equation.

$$[(30 - c)(30 + c)]\left(\frac{40}{30 - c}\right) = [(30 - c)(30 + c)]\left(\frac{80}{30 + c}\right)$$ Multiply each side by the LCD to clear the fractions.

$$40(30 + c) = 80(30 - c)$$ Simplify.
$$1200 + 40c = 2400 - 80c$$ Use distributive property.
$$1200 + 120c = 2400$$ Add 80c to each side.
$$120c = 1200$$ Subtract 1200 from each side.
$$c = 10$$ Divide each side by 120.

▶ The rate of the stream's current is 10 miles per hour.

Check ✓

Travel time downstream: $\dfrac{80}{30 + c} = \dfrac{80}{30 + 10} = \dfrac{80}{40} = 2$ hours

Travel time upstream: $\dfrac{40}{30 - c} = \dfrac{40}{30 - 10} = \dfrac{40}{20} = 2$ hours

Checkpoint ✔ **Rate Problems**

1. A plane flying at maximum speed can fly 660 kilometers with a tailwind (a wind blowing in the same direction as the plane) of 80 kilometers per hour. In the same amount of time, the plane can fly 540 kilometers against the same wind. What is the maximum speed of the plane when there is no wind?

SUMMARIZING KEY IDEAS

To solve problems involving rate, time, and distance, you can use the formula
distance = rate × time.

Exercises **Rate Problems**

1. To get to his grandmother's house, Fred must drive 140 miles on a freeway and then 20 miles along a country road. If the trip takes 3 hours, and he can travel two times faster on the freeway than on the country road, how fast does he travel on the freeway?

2. A plane can fly 340 miles per hour in still air. Flying with the wind, the plane flies 1420 miles in the same amount of time that it requires to fly 1300 miles against the wind. Find the rate of the wind.

3. It took Maggie a total of 5 hours to drive 70 miles to the airport and then fly 1800 miles to a city in Mexico. If the plane rate is 8 times faster than her car, how fast did she drive?

4. Susan rode her bike 6 miles to her friend's house and returned on foot. Her rate on the bicycle was 5 times her rate on foot. She spent 2 hours traveling to and from her friend's house. Find her rate of walking.

5. It took Rhoda the same time to drive 275 miles as it took Maureen to drive 240 miles. If Rhoda's rate was 7 miles per hour faster than Maureen's rate, how fast did each person drive?

6. Philip left home at 2:15 P.M. After driving 72 miles, he ran out of gas. He walked 2 miles to a gas station, where he arrived at 4:15 P.M. If he drove 12 times faster than he walked, how fast did he walk?

TEXTBOOK LINK

On pages S78–S80, you used the $d = rt$ formula to solve problems involving rate, time, and distance. In your textbook, Lesson 3.8, you will solve problems involving rates and ratios and in Lessons 11.6 and 11.7 you will solve rational equations for missing values.

7. The rate of a jet plane exceeds twice the rate of a cargo plane by 100 miles per hour. The jet can fly 1800 miles in the same time that the cargo plane can fly 750 miles. Use the table to find each rate.

	Distance (mi)	Rate (mi/h)	Time (h)
Jet plane	1800	$2x + 100$	?
Cargo plane	750	x	?

TEXTBOOK REFERENCES
Lessons 11.6, 11.7

KEY WORDS
• work problem

> **STUDY TIP**
> To solve problems where two or more people work together at different rates, find the part of the job that each person completes and set up an equation. The sum of the fractional parts of work done by each person equals 1, the whole job.

✏️ **My Notes**

Work Problems

Work is related to rate and time. The solution to work problems is based on the amount of work done per unit of time.

$$\boxed{\text{rate of work}} \cdot \boxed{\text{time worked}} = \boxed{\text{part of job done}}$$

Example 1

Rosa can mow the lawn in 20 minutes using a power mower. Her brother, Fidel, can mow the same lawn in 30 minutes using a hand mower. If they work together, how long will it take them to complete the job?

▶ **Solution**

Use a table to organize the data. Let $x =$ the number of minutes required to complete the job when Rosa and Fidel work together.

Worker	Rate of work (part of job done per minute)	Time worked (minutes)	Part of job done
Rosa	$\dfrac{1}{20}$	x	$x\left(\dfrac{1}{20}\right)$ or $\dfrac{x}{20}$
Fidel	$\dfrac{1}{30}$	x	$x\left(\dfrac{1}{30}\right)$ or $\dfrac{x}{30}$

$$\frac{x}{20} + \frac{x}{30} = 1 \qquad \text{Rosa's part + Fidel's part = whole job}$$

$$60\left(\frac{x}{20} + \frac{x}{30}\right) = 60(1) \qquad \text{Multiply by the LCD, 60.}$$

$$\overset{3}{\cancel{60}}\left(\frac{x}{\underset{1}{\cancel{20}}}\right) + \overset{2}{\cancel{60}}\left(\frac{x}{\underset{1}{\cancel{30}}}\right) = 60(1) \qquad \text{Use distributive property.}$$

$$3x + 2x = 60 \qquad \text{Simplify.}$$

$$5x = 60 \qquad \text{Combine like terms.}$$

$$x = 12 \qquad \text{Divide each side by 5.}$$

▶ Working together, it takes 12 minutes to mow the whole lawn.

Example 2

The larger of two pipes can fill a tank twice as fast as the smaller. Together, the two pipes require 20 minutes to fill the tank. Find the number of minutes required for the larger pipe to fill the tank.

▶ **Solution**

Let x = number of minutes needed for the larger pipe to fill the tank. Then $2x$ = number of minutes needed for the smaller pipe to fill the tank.

Size of pipe	Rate of work (part of job done per minute)	Time worked (minutes)	Part of job done
Larger pipe	$\dfrac{1}{x}$	20	$20\left(\dfrac{1}{x}\right)$ or $\dfrac{20}{x}$
Smaller pipe	$\dfrac{1}{2x}$	20	$20\left(\dfrac{1}{2x}\right)$ or $\dfrac{20}{2x}$

The sum of the fractional parts filled by each pipe is 1.

$$\frac{20}{x} + \frac{20}{2x} = 1 \qquad \text{Write equation.}$$

$$2x\left(\frac{20}{x} + \frac{20}{2x}\right) = 2x(1) \qquad \text{Multiply by the LCD, } 2x.$$

$$2x\frac{20}{x} + 2x\left(\frac{20}{2x}\right) = 2x(1) \qquad \text{Use distributive property.}$$

$$40 + 20 = 2x \qquad \text{Simplify.}$$

$$60 = 2x \qquad \text{Simplify.}$$

$$30 = x \qquad \text{Divide each side by 2.}$$

▶ The larger pipe can fill the tank in 30 minutes.

Check ✓

In 20 minutes, the larger pipe fills $\dfrac{20}{30}$, or $\dfrac{2}{3}$, of the tank. The smaller pipe fills $\dfrac{20}{60}$, or $\dfrac{1}{3}$, of the tank. $\dfrac{2}{3} + \dfrac{1}{3} = \dfrac{3}{3}$, or the whole tank. ✓

Checkpoint ✓ **Work Problems**

1. Emma can paint a fence in 2 hours. Her son Danny can paint the fence in 6 hours. Emma painted alone for 1 hour and stopped working. How many hours would Danny need to finish the job?

Summarizing Key Ideas

The solution to work problems is based on the amount of work done per unit of time. To solve problems where two or more people work together at different rates, find the part of the job each person completes and set up an appropriate equation.

Exercises Work Problems

1. Mr. Ottavino can paint the fence around his house in 3 hours. His son needs 6 hours to do the job. How many hours would it take them to do the job if they worked together?

2. Diane can weave baskets twice as fast as Sheena. Working together, they can fill a gift shop's order for baskets in 12 hours. How long will it take Diane if she works alone?

3. A school computer system can receive and process student grades from two scanners simultaneously. One scanner requires 45 minutes to read all of the grades. The other can do the same job in 30 minutes. Together, how long will it take them to read all of the grades?

4. An old machine requires three times as many hours to complete a job as a new machine. When both machines work together, they require 9 hours to complete a job. How many hours would it take the new machine to finish the job operating alone?

5. A new printing machine can complete a job in 6 hours. It takes an old machine 16 hours. If 3 new machines and 4 old machines are used to do the job, how many hours will be required to finish it?

6. A printing press can print an edition of a newspaper in 4 hours. After the press has been at work for 1 hour, another press also starts to print the edition and, together, both presses require 1 more hour to finish the job. Use the table to find the rate of work for the second press. How long would it take the second press to print the edition alone?

Printing press	Rate of work (part of job done per hour)	Time worked (hours)	Part of job done
Press #1	$\frac{1}{4}$	2	$\frac{1}{2}$
Press #2	x	1	$\frac{1}{2}$

Textbook Link

On pages S81–S83, you used rational equations to solve work problems. In your textbook, Lessons 11.6 and 11.7, you will add and subtract rational expressions and use a verbal model to help you solve problems involving rational equations.

7. A small document shredder requires $2\frac{1}{2}$ hours to shred a large pile of old documents. A larger shredder can shred all of the documents in 30 minutes. If both shredders are used, how long will it take to shred all of the documents?

(*Hint:* $2\frac{1}{2}$ hours = 150 minutes)

TEXTBOOK REFERENCES
Lessons 7.4, 11.7

KEY WORDS

• mixture problem

Percent Mixture Problems

A type of problem that chemists and pharmacists can encounter involves changing the concentration of a solution or other mixture. In such problems, the amount of a particular ingredient in the solution or mixture is often expressed as a percent of the total. You can use a table to help you solve a mixture problem.

Example 1

A chemist adds 3 liters of acid to 15 liters of an acid solution that is 40% acid. What is the percent of the new solution that is acid?

▶ **Solution**

Let x = the percent of the new solution that is acid.

	% acid	Total amount (liters)	Amount of acid (liters)
Start with	40%	15	0.40(15) = 6
Add acid	100%	3	1.00(3) = 3
Finish with	x	18	x(18)

> **STUDY TIP**
> To convert a percent to a decimal, move the decimal point two places to the left. For example, 40% = 0.40.

The amount of acid in the original solution plus the amount of acid added equals the amount of acid in the new solution.

$$0.40(15) + 1.00(3) = x(18) \qquad \text{Write equation.}$$
$$6 + 3 = 18x \qquad \text{Simplify.}$$
$$9 = 18x \qquad \text{Simplify.}$$
$$\frac{1}{2} = x \qquad \text{Divide each side by 18.}$$

▶ The percent of the new solution that is acid will be $\frac{1}{2}$ = 0.50 = 50%.

Example 2

A 50 milliliter solution of acid and water contains 25% acid. How much water do you need to add in order to make a 10% acid solution?

▶ **Solution**

Let x = the number of milliliters of water to be added.

	% acid	Total amount (mL)	Amount of acid (mL)
Start with	25%	50	0.25(50)
Add water	0%	x	0
Finish with	10%	$50 + x$	$0.10(50 + x)$

You can clear the equation of decimals by multiplying each side of the equation by 100.

Since the number of milliliters of acid in the solution stays the same, the first and third entries of the last column are equal.

$0.25(50) = 0.10(50 + x)$	Write equation.
$100(0.25)(50) = 100(0.10)(50 + x)$	Multiply each side by 100.
$25(50) = 10(50 + x)$	Simplify.
$1250 = 500 + 10x$	Use distributive property.
$750 = 10x$	Subtract 500 from each side.
$75 = x$	Divide each side by 10.

▶ 75 milliliters of water must be added to make a 10% acid solution.

Check ✓

The amount of acid should be equal to ten percent of the total volume of the new solution.

$$0.25(50) \stackrel{?}{=} 0.10(50 + x)$$
$$12.5 \stackrel{?}{=} 0.10(50 + 75)$$
$$12.5 \stackrel{?}{=} 0.10(125)$$
$$12.5 = 12.5 ✓$$

My Notes

Checkpoint ✓ *Percent Mixture Problems*

1. A chemist has two solutions, one 55% acid and the other 30% acid. How many grams of each solution must be mixed to produce 60 grams of a solution that is 35% acid?

Exercises Percent Mixture Problems

1. How many milliliters of water would you add to 75 milliliters of a 15% acid solution to make a 10% acid solution?

2. If you wish to increase the percent of acid in 50 milliliters of a 10% acid solution in water to 25% acid, how much pure acid must you add?

3. You add 1 quart of orange juice concentrate to 4 quarts of a solution that is 20% concentrate in water. What is the percent of concentrate in your new solution?

4. How many ounces of meat should be added to 125 ounces of a 45% mixture of cereal in meat in order to produce a pet food mixture that is 30% cereal?

5. In order to increase the amount of salt in a 40 gram mixture from 5% to 20% salt, how much salt must be added?

6. If 3 quarts of pure antifreeze are added to 9 quarts of a 40% antifreeze solution, what is the percent of antifreeze in the new solution?

7. An auto mechanic adds 25 milliliters of water to 125 milliliters of a 15% solution of antifreeze in water. What is the percent of antifreeze in the new solution?

8. You add 50 grams of ground Costa Rican coffee to 200 grams of a ground mixture that contains 5% Costa Rican coffee. What is the percent of Costa Rican coffee in the new mixture?

9. A chemist adds 2 liters of water to 5 liters of a 42% nitric acid solution. What is the percent of acid in the new solution?

10. How many gallons of milk that is $3\frac{1}{2}$% milk fat must be added to 80 gallons of milk that is 1% milk fat to make milk that is 2% milk fat?

TEXTBOOK LINK

On pages S84–S86, you wrote and solved equations for percent mixture problems. In your textbook, Lesson 11.7, you will solve more mixture problems using rational equations. In Lesson 7.4, you will use systems of linear equations to solve a variety of mixture problems.

Topic Review *Problem Solving*

These exercises will help you check that you can apply algebraic techniques to solve rate problems, work problems, and percent mixture problems. If you have any questions about problem solving, be sure to get them answered before going on to the next section.

1. James drives 5 miles per hour slower than Ken. If James travels 100 miles in the same time that Ken travels 110 miles, find each rate.

2. A chemist mixes 10 liters of a solution that is 30% acid with 30 liters of a solution that is 50% acid. What is the percent of acid in the new mixture?

3. Your friend can address envelopes twice as fast as you can. Together, you and your friend address all of the invitation envelopes in 3 hours. How long would it take your friend to address the envelopes alone?

4. How many milliliters of acid must be added to 50 milliliters of a 10% acid solution in order to produce a 20% acid solution?

5. On Saturday, Earl rode his bike for 2 hours longer than Ted. Earl traveled 135 kilometers and Ted traveled 81 kilometers. If they both averaged the same rate of speed, how long did Earl ride?

6. Greg lives 25 miles from his school. One day he takes the bus to school and returns home by car. His total travel time is $1\frac{1}{2}$ hours.

    If the car travels at twice the rate of the bus, find the average rates of the bus and the car.

7. Cerise can paint the living room in 2 hours and Kyle can paint it in 3 hours. If they work together, how long will it take them to complete the job?

8. You have 15 liters of a 24% acid solution. How much water should you add in order to dilute the solution so that it is 8% acid?

9. Karen traveled 640 miles by boat and returned by plane. She spent 42 hours traveling on the trip. If the rate of the plane was 20 times faster than the rate of the boat, find each rate.

My Review Questions

Students understand the concepts of a relation and a function, determine whether a given relation defines a function, and give pertinent information about given relations and functions.

TEXTBOOK REFERENCES
Lessons 1.8, 4.8

KEY WORDS
- relation
- function
- domain
- range
- vertical line test
- x-y notation
- function notation

Functions and Relations

IDENTIFYING FUNCTIONS

Any set of ordered pairs is called a **relation**. A **function** is a relation in which each input has exactly one output. The **domain** of a function is the collection of all possible input values. The **range** of a function is the collection of all possible output values.

Example 1

Determine whether the relation is a function. If it is a function, give the domain and range.

a. $(0, -1), (2, -2), (1, -1),$
$(1, 5), (-3, 4)$

b.

Time (hours)	Temperature (°C)
1	15
2	16
3	17
4	19
5	17

▶ **Solution**

a. The relation is not a function because the input 1 has two outputs: -1 and 5.

b. The relation is a function. There is exactly one temperature for each time. The domain is 1, 2, 3, 4, and 5. The range is 15, 16, 17, and 19.

Checkpoint ✓ *Identify Functions*

Determine whether the relation is a function. If it is a function, give the domain and range.

1. $(-3, -2), (-2, -1), (-1, 0),$
$(0, 1), (1, 2)$

2. Input   Output

My Notes

VERTICAL LINE TEST

A method for determining whether a graph represents a function involves drawing vertical lines through the graph of the relation.

> **VERTICAL LINE TEST**
>
> A graph is a function if any vertical line intersects the graph at no more than one point.

Example 2

Use the vertical line test to determine whether the graph represents a function. Explain your reasoning.

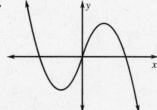

a. **b.**

▶ *Solution*

a. The graph represents a function since no vertical line can intersect the graph more than once.

b. It is possible to draw a vertical line that intersects the graph twice. So, this graph does *not* represent a function.

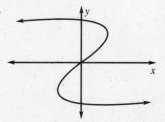

Checkpoint ✓ *Vertical Line Test*

Use the vertical line test to determine whether the graph represents a function. Explain your reasoning.

3.  **4.** 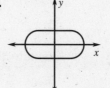 **5.**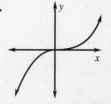

▶ **STUDY TIP**
The symbol $f(x)$ does not mean f times x, rather it refers to the value of f at x. For example, $f(4)$ refers to the value of the function f when evaluated at 4.

EVALUATING FUNCTIONS

A function can be expressed using *x-y notation* or *function notation*:

x-y notation: $y = 2x - 7$

function notation: $f(x) = 2x - 7$

You can evaluate a function for a given value by substituting the given value for the variable and simplifying.

Example 3

Evaluate the function $y = 9x - 10$ when $x = -4$.

▶ **Solution**

$y = 9x - 10$	Write original function.
$y = 9(-4) - 10$	Substitute -4 for x.
$y = -46$	Simplify.

▶ When $x = -4$, $y = -46$.

▶ **STUDY TIP**
You don't have to use f to name a function. You can use other letters such as g and h.

Example 4

Evaluate the function $f(x) = 2x - 3$ for the given value of x.

  a. $x = 0$ **b.** $x = 2$ **c.** $x = 10$ **d.** $x = -1$

▶ **Solution**

  a. $f(0) = 2(0) - 3 = 0 - 3 = -3$

  b. $f(2) = 2(2) - 3 = 4 - 3 = 1$

  c. $f(10) = 2(10) - 3 = 20 - 3 = 17$

  d. $f(-1) = 2(-1) - 3 = -2 - 3 = -5$

Checkpoint ✓ **Evaluate Functions**

Evaluate the function when $x = 6$.

  6. $y = x - 11$ **7.** $y = 3x + 1$ **8.** $y = -4x + 4$

Evaluate the function $f(x) = x - 1$ for the given value of x.

  9. $x = 2$ **10.** $x = -2$ **11.** $x = 0$ **12.** $x = 1$

A **theorem** is a general conclusion that is shown to be true by using axioms, definitions, given facts, and other proven theorems. The reasoning that takes you from the hypothesis (the given statement) to the conclusion (the final statement) is called a **direct proof.**

Example 1

Prove that for real numbers *a* and *b*, $a + ab = a(b + 1)$.

▶ *Solution*

Statements	Reasons
1. $a + ab = a \cdot 1 + ab$	**1.** Multiplicative identity
2. $\quad = a(1 + b)$	**2.** Distributive property
3. $\quad = a(b + 1)$	**3.** Commutative prop. of addition

Example 2

Prove that for real numbers *a* and *b*, $-(a - b) = b - a$.

▶ *Solution*

Statements	Reasons
1. $-(a - b) = -[a + (-b)]$	**1.** Subtraction property
2. $\quad = -1[a + (-b)]$	**2.** Property of negative one
3. $\quad = (-1)a + (-1)(-b)$	**3.** Distributive property
4. $\quad = (-a) + (b)$	**4.** Property of negative one
5. $\quad = b + (-a)$	**5.** Commutative prop. of addition
6. $\quad = b - a$	**6.** Subtraction property

Checkpoint ✔ **Prove a Theorem**

3. Copy and complete the proof of the following statement.

For real numbers a *and* b, $a + b + (-b) = a$.

Statements	Reasons
1. $(a + b) + (-b) = a + [b + (-b)]$	**1.** ___?___
2. $\quad = a + $ ___?___	**2.** Inverse prop. of addition
3. $\quad = a$	**3.** ___?___

My Notes

Another way to prove a statement is by assuming the statement is false. If this assumption leads to an impossibility, then the original statement is true. This type of proof is called an **indirect proof.**

Example 3

Prove that for all real numbers *a*, *b*, and *c* > 0, if *a* > *b* then *ac* > *bc*.

▶ **Solution**

Assume that *ac* is *not* greater than *bc*, which means $ac \leq bc$.

Statements	Reasons
1. $a > b$ and $c > 0$	**1.** Given
2. Suppose $ac \leq bc$.	**2.** Supposition
3. $ac \cdot \dfrac{1}{c} \leq bc \cdot \dfrac{1}{c}$	**3.** Multiplication prop. of inequalities
4. $a\left(c \cdot \dfrac{1}{c}\right) \leq b\left(c \cdot \dfrac{1}{c}\right)$	**4.** Associative prop. of multiplication
5. $a(1) \leq b(1)$	**5.** Multiplicative inverse
6. $a \leq b$	**6.** Multiplicative identity

▶ But $a \leq b$ contradicts the given that $a > b$. Therefore the supposition $ac \leq bc$ is false, so $ac > bc$.

> **STUDY TIP**
> When both sides of an inequality are multiplied (or divided) by a negative number, the direction of the inequality sign is reversed. In step 3 of Example 3, because it is given that $c > 0$, the inequality sign is not reversed.

If a person makes a statement that he or she supposes to be true, just one *counterexample* proves that the statement is false.

Example 4

Show that the following statement is false by finding a counterexample.

If a *and* b *are real numbers, then* $-a + b = -b + a$.

▶ **Solution**

Let $a = 1$ and $b = 2$. Then $-a + b = -1 + 2 = 1$, but $-b + a = -2 + 1 = -1$. Since $1 \neq -1$, the counterexample $a = 1$ and $b = 2$ shows that the statement made above is false.

Checkpoint ✓ **Use Logical Reasoning**

4. Show that the statement below is false by finding a counterexample.

If a, b, *and* c *are real numbers, then* $2a + 3b = 5 + a + b$.

SUMMARIZING KEY IDEAS

To prove statements about real numbers, use postulates, definitions, given facts, and other proved theorems.

Exercises *Logical Reasoning: Proof*

State the basic axiom of algebra that is represented.

1. If $j = k$, then $k = j$.   **2.** $0 + p = 0$ **3.** $x \cdot 0 = 0$

4. $3 - 2 = 3 + (-2)$ **5.** $6 \cdot \dfrac{1}{6} = 1$ **6.** If $c = d$, then $ca = da$.

7. Copy and complete the proof of the following statement.

For all real numbers x and y, $-(x - y) + (y - x) = 2(y - x)$.

Statements	Reasons
1. $-(x - y) + (y - x)$ $\quad = -[x + (-y)] + [y + (-x)]$	**1.** Subtraction property
2. $\quad = -1[x + (-y)] + [y + (-x)]$	**2.** _?_
3. $\quad = (-1)(x) + (-1)(-y) + [y + (-x)]$	**3.** _?_
4. $\quad = -x + y + [y + (-x)]$	**4.** _?_
5. $\quad = y + y + (-x) + (-x)$	**5.** Commutative prop. of addition
6. $\quad = 2y + (-2x)$	**6.** Combine like terms.
7. $\quad = 2[y + (-x)]$	**7.** _?_
8. $\quad = 2(y - x)$	**8.** _?_

Prove the theorem.

8. If a and b are real numbers and $a \neq 0$, then $a \cdot \dfrac{b}{a} = b$.

9. If a, b, and c are real numbers, then $ab - bc = b(a - c)$.

Use an indirect proof to prove the theorem.

10. If a, b and c are real numbers and $a - c > b - c$, then $a > b$.

11. If a and b are positive integers and $2a + b$ is divisible by 2, then b must be divisible by 2. (*Hint:* Assume that b is odd, meaning $b = 2n + 1$ for some positive integer n.)

TEXTBOOK LINK

On pages S92–S95, you used axioms and logical reasoning to prove and disprove mathematical statements. You will learn more about proving theorems in Lesson 12.9 in your textbook.

Show that the statement is false by finding a counterexample.

12. If a, b, and c are real numbers and $a - b > a - c$, then $b > c$.

13. If a and b are real numbers and $a(a - b) = 0$, then $a = b$.

14. If a, b, and c are real numbers, then $\dfrac{c}{a - b} = \dfrac{c}{a} - \dfrac{c}{b}$.

Part 3 Special Topics

Getting Started with Word Problems

TEXTBOOK REFERENCES
Lesson 1.6

CA STANDARDS
5.0

If you have difficulty getting started with word problems, it may help you to write a word equation before trying to write a verbal model.

Example 1 *Write a Word Equation*

Jesse collects both baseball and football cards. He has twice as many football cards as baseball cards. Altogether, he has 195 cards. How many baseball cards does he have?

▶ *Solution*

Begin by writing a simple word equation that describes the main idea in the problem.

baseball cards plus football cards = 195

- Since the problem asks you to find how many baseball cards Jesse has, you can let a variable, say b (for baseball), represent that number.

- Jesse has twice as many football cards as baseball cards, so that number can be represented by the variable expression $2b$.

- Now use your word equation to write an equation using variables.

Word equation	baseball cards	plus	football cards	=	195
Algebraic model	b	+	$2b$	=	195

$$b + 2b = 195$$
$$3b = 195$$
$$b = 65$$

▶ *Answer* Jesse has 65 baseball cards.

In some cases, you may find it helpful to draw a diagram.

Example 2 **Draw a Diagram**

Raymond and Elsa were saving to buy computer software. When they began saving, Elsa had $42 and Raymond had $17. They agreed to help their parents with yard work to earn extra money. Elsa and Raymond were paid the same amount. After being paid, Elsa had twice as much as Raymond. How much was each person paid?

▶ **Solution**

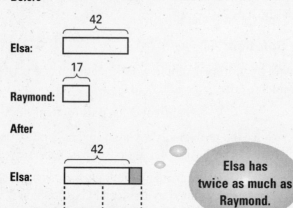

Before

42

Elsa:

17

Raymond:

After

42

Elsa:

Elsa has twice as much as Raymond.

Raymond:

17   17

METHOD 1 Compare the diagrams: $42 - \$17 - \$17 = \$42 - \$34 = \$8$

METHOD 2 Use the diagrams to write an equation. Let x represent the shaded block, the amount each person was paid.

$$42 = 34 + x \qquad x = 8$$

▶ **Answer** Elsa and Raymond were each paid $8.

Checkpoint ✔ **Getting Started with Word Problems**

You pay $3.40 for a lunch consisting of a roast beef sandwich and a glass of milk. The sandwich costs three times as much as the milk.

1. Write a word equation that describes the situation.

2. Draw a diagram that describes the situation.

3. Find the cost of a glass of milk.

Write a word equation that describes the situation.

1. There are ten more oak trees than maple trees in the town park. The total number of oaks and maples is 58. How many maple trees are there?

Draw a diagram that describes the situation.

2. You spent a total of six hours preparing a history report. You spent twice as long researching the report as you did writing and typing it. How long did you spend researching?

3. Andre has eight more CDs than Zita has. If he gets four more, he will have twice as many CDs as she has. How many CDs does each person have?

4. Orla is 10 years older than her brother. In five years, she will be twice his age. How old is Orla? How old is her brother?

5. Sixto has a collection of glass angels. The shelf holding the angels was knocked over and one-third of them were broken. Sixto has 12 statues left. How many did he have before?

6. Laura scored four times as many two-point goals as three-point goals in a basketball game. She scored a total of 33 points. How many two-point goals did she score?

7. Bradley and Tameeka collected money for a local shelter. Together they collected $84. Bradley collected $8 more than Tameeka. How much did each person collect?

8. A tennis camp offered two sessions. At the beginning of the season there were 65 people enrolled in the early session and 25 enrolled in the late session. After the first week, some of the people in the early session switched to the late session. Afterward, there were twice as many people in the early session as in the late session. How many people were in each session after the first week?

9. Miguel is planning a birthday party at a pizza parlor. It will cost $30 for 12 people with an additional $3 per person for any extra guests. If he has $45 to spend, how many guests can he invite?

TEXTBOOK REFERENCES

Lesson 4.5

KEY WORDS

• slope
• rise
• run

CA STANDARDS

24.0, 25.0

To find the **slope** of a nonvertical line, first choose two points $P(x_1, y_1)$ and $Q(x_2, y_2)$ on the line. Then calculate the ratio of the vertical **rise** to the horizontal **run** from P to Q.

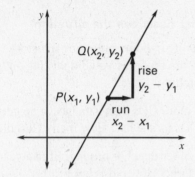

$$\text{slope} = \frac{\text{rise}}{\text{run}} = \frac{y_2 - y_1}{x_2 - x_1}$$

You might ask, "Does the value of the slope depend on the two points used to calculate it?" The activity will help you answer this question.

ACTIVITY INVESTIGATING THE SLOPE OF A LINE

For Steps 1–3, use the graph below.

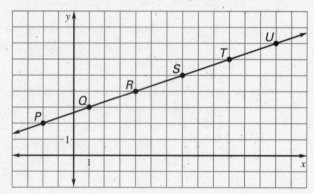

❶ Use the given pair of points to calculate the slope of the line.

   a. P and Q **b.** P and R **c.** P and T

   d. Q and S **e.** Q and U **f.** R and T

❷ What do you notice about the slopes you found in Step 1?

❸ Make a conjecture about whether the slope of a line depends on the points used to calculate it.

Your work in the activity *suggests* that you get the same value for a line's slope no matter which two points you use to calculate it. However, you need a *proof* to be certain that this result is true.

The next example uses concepts from geometry to prove that the slope of a line is constant. To understand the proof, you'll need to recall the following results from a previous course.

- **Corresponding angles postulate:** If two parallel lines are cut by a third line called a *transversal*, then corresponding angles have equal measure.

- **Angle-angle similarity postulate:** If two angles of one triangle have the same measures as two angles of another triangle, then the triangles are similar.

Example 1 *A Geometric Proof That a Line's Slope Is Constant*

Prove that the slope of a nonvertical line does not depend on which two points are used to calculate it.

▶ *Solution*

Let P, Q, R, and S be any four points on the line. It is enough to prove that the slope obtained using P and Q equals the slope obtained using R and S.

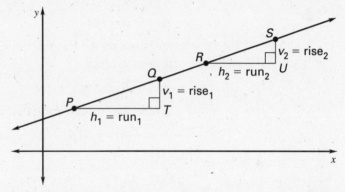

In the diagram, the line is a transversal of the parallel segments $\overline{PT}$ and $\overline{RU}$. Since $\angle P$ and $\angle R$ are corresponding angles, $m\angle P = m\angle R$ by the corresponding angles postulate. Also, $m\angle T = m\angle U = 90°$, so $\triangle PQT$ and $\triangle RSU$ are similar by the angle-angle similarity postulate. Therefore:

▶**STUDY TIP**
Recall that if two figures are similar, then the ratios of the lengths of corresponding sides are equal.

$$\frac{v_1}{v_2} = \frac{h_1}{h_2}$$ Definition of similar triangles

$$v_1 \cdot h_2 = h_1 \cdot v_2$$ Use cross products property.

$$\frac{v_1 \cdot h_2}{h_1 \cdot h_2} = \frac{h_1 \cdot v_2}{h_1 \cdot h_2}$$ Divide each side by $h_1 \cdot h_2$.

$$\frac{v_1}{h_1} = \frac{v_2}{h_2}$$ Simplify.

$$\begin{matrix} \text{Slope using} \\ P \text{ and } Q \end{matrix} = \begin{matrix} \text{Slope using} \\ R \text{ and } S \end{matrix}$$ The slope does not depend on the points used to find it.

1. In Example 1, how do you know that $\overline{PT}$ and $\overline{RU}$ are parallel?

2. The proof in Example 1 is based on a line that slopes *upward* from left to right. Modify the proof to account for the case where the line slopes *downward* from left to right.

It is not necessary to use ideas from geometry to prove that the slope of a line is constant. The following example gives a purely algebraic proof of this result.

Example 2 *An Algebraic Proof That a Line's Slope Is Constant*

Prove that the slope of a nonvertical line does not depend on which two points are used to calculate it.

▶ **Solution**

A nonvertical line has an equation of the form $Ax + By = C$ where $B \neq 0$. Let $P(x_1, y_1)$ and $Q(x_2, y_2)$ be any two different points on the line. Then:

$$Ax_1 + By_1 = C \qquad \text{Equation 1: } P \text{ lies on the line.}$$
$$Ax_2 + By_2 = C \qquad \text{Equation 2: } Q \text{ lies on the line.}$$

Since the expressions on the left sides of Equations 1 and 2 both equal C, they are also equal to each other.

$$Ax_1 + By_1 = Ax_2 + By_2 \qquad \text{Equate left sides of Equations 1 and 2.}$$
$$-By_2 + By_1 = Ax_2 - Ax_1 \qquad \text{Add } -Ax_1 - By_2 \text{ to each side.}$$
$$-B(y_2 - y_1) = A(x_2 - x_1) \qquad \text{Factor.}$$
$$\frac{-B(y_2 - y_1)}{-B(x_2 - x_1)} = \frac{A(x_2 - x_1)}{-B(x_2 - x_1)} \qquad \text{Divide each side by } -B(x_2 - x_1).$$
$$\frac{y_2 - y_1}{x_2 - x_1} = -\frac{A}{B} \qquad \text{Simplify.}$$

$$\text{Slope using } P \text{ and } Q = -\frac{A}{B} \qquad \qquad \frac{y_2 - y_1}{x_2 - x_1} = \text{Slope using } P \text{ and } Q$$

The slope obtained using points P and Q is a constant that does not depend on the coordinates of P or Q. Therefore, any two points on the line will produce the same slope.

Checkpoint ✓ **Understanding the Algebraic Proof**

3. Using the proof in Example 2, find the slope of the line with equation $3x + 5y = 9$ simply by looking at the constants in the equation.

Find the slope of the line through the points.

1. $(0, 0), (1, 3)$ **2.** $(-4, 5), (-2, 1)$ **3.** $(-5, -6), (4, -6)$

4. $(7, 9), (-5, 1)$ **5.** $(-13, 10), (-21, 0)$ **6.** $(-1, 0.5), (0.5, 8)$

Tell whether the three points lie on the same line. Explain your answer.

7. $(1, 5), (2, 7), (4, 11)$ **8.** $(-3, 13), (0, 1), (5, -14)$

9. $(-4, -1), (-1, 1), (11, 17)$ **10.** $(-6, -28), (2, 0), (12, 35)$

In Exercises 11 and 12, find the value of the variable for which the three points lie on the same line.

11. $(0, -4), (2, 2), (3, y)$ **12.** $(x, 13), (2, 1), (7, -9)$

13. The slope of the line with equation $Ax + 2y = 11$ is 4. Use the proof given in Example 2 on page T6 to find the value of A.

14. GEOMETRY The vertices of a square are $P(3, 5)$, $Q(11, 3)$, $R(9, -5)$, and $S(1, -3)$. Use the concept of slope to show that $T(6, 0)$ lies on the diagonal joining P and R, and on the diagonal joining Q and S.

15. MATHEMATICAL REASONING The equation of any nonvertical line can be written in the form $y = mx + b$ where m and b are constants. Susan wants to prove that the slope of a line with equation $y = mx + b$ is constant and equal to m. Complete Susan's proof.

> Let $P(x_1, y_1)$ and $Q(x_2, y_2)$ be any two points on the line $y = mx + b$.
>
> $$\text{Slope} = \frac{y_2 - y_1}{x_2 - x_1}$$
>
> $$= \frac{(mx_2 + b) - (mx_1 + b)}{x_2 - x_1}$$

16. CHALLENGE Recall that two lines are parallel if they do not intersect. Prove that if the slopes of two distinct nonvertical lines are equal, then the lines are parallel.

(*Hint:* Prove the logically equivalent statement "If two distinct lines intersect, then their slopes are not equal." To do this, write the equations of the lines as $y = m_1 x + b_1$ and $y = m_2 x + b_2$ where m_1 and m_2 are the slopes. Assume the lines intersect at the point (x_0, y_0). Then

$$y_0 = m_1 x_0 + b_1 \quad \text{and} \quad y_0 = m_2 x_0 + b_2,$$

so that $m_1 x_0 + b_1 = m_2 x_0 + b_2$. Solve this last equation for x_0, and explain why the result shows that $m_1 \neq m_2$.)

TEXTBOOK REFERENCES

Lessons 9.1, 11.6

KEY WORDS

- square root
- Babylonian method
- iteration
- rational expression
- rational number
- irrational number
- even number
- odd number
- prime number
- cube root

CA STANDARDS

1.0, 2.0, 13.0, 25.0

APPROXIMATING SQUARE ROOTS

Recall that the positive square root of a positive number a, written $\sqrt{a}$, is a number b such that $b^2 = a$. For example, $\sqrt{1} = 1$ and $\sqrt{4} = 2$. Finding or approximating square roots of numbers other than small perfect squares like 1 and 4 can be challenging. One way is to make and examine lists.

Example 1   Approximating a Square Root Using Lists

Approximate $\sqrt{2}$ **(a)** to the nearest tenth and **(b)** to the nearest hundredth.

▶ **Solution**

a. Because 2 is between 1 and 4, you can reason that $\sqrt{2}$ is between $\sqrt{1}$ and $\sqrt{4}$, or between 1 and 2. Make and examine a list of squares of the numbers 1.0, 1.1, ... , 1.9, 2.0. (Note that you can terminate the list as soon as you find a square greater than 2.)

$$1.0^2 = 1.0$$
$$1.1^2 = 1.21$$
$$1.2^2 = 1.44$$
$$1.3^2 = 1.69$$
$$1.4^2 = 1.96$$
$$1.5^2 = 2.25$$ $\leftarrow$ $\sqrt{2}$ lies between 1.4 and 1.5.

▶ **Answer** Because $1.4^2 = 1.96$ is much closer to 2 than $1.5^2 = 2.25$ is, $\sqrt{2} \approx 1.4$ to the nearest tenth.

b. Because $\sqrt{2}$ is between 1.4 and 1.5, make and examine a list of squares of the numbers 1.40, 1.41, ... , 1.49, 1.50. (As before, you can terminate the list as soon as you find a square greater than 2.)

$$1.40^2 = 1.96$$
$$1.41^2 = 1.9881$$
$$1.42^2 = 2.0164$$ $\leftarrow$ $\sqrt{2}$ lies between 1.41 and 1.42.

While $1.41^2 = 1.9881$ is closer to 2 than $1.42^2 = 2.0164$ is, the differences are nearly equal, so to decide whether 1.41 or 1.42 gives $\sqrt{2}$ to the nearest hundredth, check 1.415 (the number that determines whether you round up or round down). Because $1.415^2 = 2.002225$, you know that $\sqrt{2}$ is less than 1.415, and you round down.

▶ **Answer** $\sqrt{2} \approx 1.41$ to the nearest hundredth.

· ·

You can continue the method shown in Example 1 to get better approximations of $\sqrt{2}$, but the work is tedious.

The following example develops an efficient method of approximating a square root. Historians believe this method was known and used by the ancient Babylonians several thousand years ago.

Example 2 *Developing the Babylonian Method*

Suppose you have a reasonable approximation of $\sqrt{2}$ and you want to improve it. Let b be the approximation, and let d be the difference between $\sqrt{2}$ and b. Estimate the value of d (in terms of b) and use it to find a better approximation of $\sqrt{2}$.

▶ **Solution**

> **▶STUDY TIP**
> An expression like $\dfrac{2-b^2}{2b}$ is called a *rational expression*. You simplify and perform arithmetic operations on rational expressions in the same ways as you do fractions.

If you could know the value of d exactly, then $b + d = \sqrt{2}$, or $(b + d)^2 = 2$. Squaring the binomial gives $b^2 + 2bd + d^2 = 2$. If b^2 is already close to 2, then d^2 is quite small and can be ignored. You can therefore write and solve the following for d:

$$b^2 + 2bd \approx 2 \qquad \text{Drop the } d^2 \text{ term.}$$

$$2bd \approx 2 - b^2 \qquad \text{Subtract } b^2 \text{ from both sides.}$$

$$d \approx \frac{2 - b^2}{2b} \qquad \text{Divide both sides by } 2b.$$

This estimate for the value of d can be used to obtain a better approximation of $\sqrt{2}$ as follows:

$$\sqrt{2} = b + d \qquad \text{Express } \sqrt{2} \text{ in terms of } b \text{ and } d.$$

$$\approx b + \frac{2 - b^2}{2b} \qquad \text{Substitute } \frac{2 - b^2}{2b} \text{ for } d.$$

$$= \frac{2b^2}{2b} + \frac{2 - b^2}{2b} \qquad \text{Get a common denominator.}$$

$$= \frac{2b^2 + 2 - b^2}{2b} \qquad \text{Add numerators.}$$

$$= \frac{b^2 + 2}{2b} \qquad \text{Simplify.}$$

By dividing the numerator and denominator of the last expression by b, you get $\dfrac{b + \dfrac{2}{b}}{2}$, which looks more complicated but is easier to interpret: To obtain a better approximation of $\sqrt{2}$ than b is, just divide 2 by b and take the average of b and the quotient. Because of this, the Babylonian method of approximating a square root is sometimes called the divide-and-average method.

. . . .

 **Using the Babylonian Method**

1. Let 1.4 be your first good approximation of $\sqrt{2}$. Use the Babylonian method to find a better approximation. That is, find the value of $\dfrac{1.4 + \dfrac{2}{1.4}}{2}$ (using a calculator). Compare this with 1.414213562, which is the value of $\sqrt{2}$ to the nearest billionth.

2. Using your calculator's memory feature, take the approximation from Exercise 1 and substitute it for b in the expression $\dfrac{b + \dfrac{2}{b}}{2}$. How does this new approximation compare with 1.414213562?

You can generalize the Babylonian method to any square root, not just $\sqrt{2}$.

BABYLONIAN METHOD FOR APPROXIMATING A SQUARE ROOT

If b is an approximixation of $\sqrt{a}$, then a better approximation is given by:

$$\frac{b + \dfrac{a}{b}}{2}$$

The new approximation can in turn be improved by substituting it for b in the expression above. This process of repeatedly substituting "old" values in an expression to obtain "new" values is called iteration and in the case of the Babylonian method can be written as

$$b_{new} = \frac{b_{old} + \dfrac{a}{b_{old}}}{2}$$

where in each iteration b_{old} takes on the value of b_{new} from the previous iteration.

PROVING SOME SQUARE ROOTS ARE IRRATIONAL

The ancient Greek mathematician Pythagoras and his followers believed that "all is number," and for them, numbers consisted solely of whole numbers (0, 1, 2, 3, . . .) and their ratios. (Today, we define a *rational* number as the ratio of two integers, such as $\dfrac{2}{3}$ and $\dfrac{-1}{5}$. The ancient Greeks, however, were familiar only with ratios of positive integers.) When the ancient Greeks encountered a number like $\sqrt{2}$ through their work with the Pythagorean theorem, it was natural for them to ask what fraction $\sqrt{2}$ is. What they found is that there is no such fraction.

The next two examples give proofs that $\sqrt{2}$ cannot be written as the ratio of two integers. Such numbers are called *irrational*.

Example 3 Proving that $\sqrt{2}$ Is Irrational

Prove that $\sqrt{2}$ cannot be written in the form $\frac{p}{q}$ where p and q are whole numbers and $q \neq 0$.

▶ **Solution**

Begin by assuming that $\sqrt{2}$ can be written as $\frac{p}{q}$ where p and q have no common factor. This assumption leads to the following:

$$\frac{p}{q} = \sqrt{2} \qquad \text{Assumption}$$

$$\frac{p^2}{q^2} = 2 \qquad \text{Square both sides.}$$

$$p^2 = 2q^2 \qquad \text{Multiply both sides by } q^2.$$

Now consider whether p is an even or an odd number. If p is odd, then $p = 2k + 1$ for some whole number k. Then $p^2 = (2k + 1)^2 = 4k^2 + 4k + 1 = 2(2k^2 + 2k) + 1$, which is also an odd number because it is 1 more than an even number. However, the fact that $p^2 = 2q^2$ tells you that p^2 is even. Since p^2 cannot be both odd and even, p itself must not be odd. Therefore, p is even.

Since p is even, you know that $p = 2k$ for some whole number k, and you can substitute $2k$ for p in the equation $p^2 = 2q^2$ as follows:

$$(2k)^2 = 2q^2 \qquad \text{Substitute } 2k \text{ for } p.$$

$$4k^2 = 2q^2 \qquad \text{Multiply.}$$

$$2k^2 = q^2 \qquad \text{Divide both sides by 2.}$$

As you did with p, you can argue that q must be even. Since both p and q are even, they have 2 as a common factor. But the assumption was that p and q have *no* common factor. Since you have arrived at a contradiction (p and q cannot both have a common factor and not have one), you know that an error has been made. Since there is no flaw in the argument after the assumption was made, the assumption itself must be wrong. Therefore, you can conclude that there are no whole numbers p and q with $\frac{p}{q} = \sqrt{2}$.

· ·

The next proof that $\sqrt{2}$ is irrational also begins by assuming that $\sqrt{2}$ is rational and shows that a contradiction results.

▶ **VOCABULARY TIP**
A whole number is even provided it be can written in the form $2k$ for some whole number k. (For instance, you can write 10 as $2 \cdot 5$, so 10 is even.) A whole number is odd provided it can be written in the form $2k + 1$ for some whole number k. (For instance, you can write 13 as $2 \cdot 6 + 1$, so 13 is odd.)

Example 4 Proving that $\sqrt{2}$ Is Irrational

Prove that $\sqrt{2}$ cannot be written in the form $\dfrac{p}{q}$ where p and q are whole numbers and $q \neq 0$.

▶ **Solution**

Begin by assuming that $\sqrt{2}$ can be written as $\dfrac{p}{q}$. As in Example 3, this assumption leads to the conclusion that $p^2 = 2q^2$. Now imagine writing p and q as the product of prime numbers. A theorem from number theory says that factoring a whole number greater than 1 into a product of primes can always be done and that there is only one such factorization. Since every prime that appears in the factorization of p appears twice as often in the factorization of p^2, each prime in the factorization of p^2 appears an even number of times. The same is true for the factorization of q^2. In the equation $p^2 = 2q^2$, however, the prime number 2 must appear an even number of times on the left side but an odd number of times on the right. In other words, you have two numbers, p^2 and $2q^2$, that are equal but have different prime factorizations. This contradiction leads you to conclude that the assumption must be wrong. Therefore, you can conclude that there are no whole numbers p and q with $\dfrac{p}{q} = \sqrt{2}$.

▶ **VOCABULARY TIP**
A *prime* number is a whole number greater than 1 that has only 1 and itself as factors. The first few prime numbers are 2, 3, 5, 7, 11, and 13.

Exercises

1. a. Make and examine lists of squares to approximate $\sqrt{3}$ to the nearest tenth and then to the nearest hundredth.

 b. Use the Babylonian method to approximate $\sqrt{3}$ to the nearest thousandth. How do you know when to stop the iteration?

2. Is $\sqrt{169}$ a rational number or an irrational number? Explain.

▶ **VOCABULARY TIP**
A *cube root* of a number a is a number b such that $b^3 = a$. A cube root of a is written $\sqrt[3]{a}$.

3. MATHEMATICAL REASONING Suppose you have a reasonable approximation of $\sqrt[3]{2}$ and you want to improve it. Let b be the approximation, and let d be the difference between $\sqrt[3]{2}$ and b. Estimate the value of d (in terms of b) and use it to find a better approximation of $\sqrt[3]{2}$. (*Hint:* $(b + d)^3 = b^3 + 3b^2d + 3bd^2 + d^3$. If b^3 is close to 2, then both d^2 and d^3 are quite small.)

4. MATHEMATICAL REASONING Prove that $\sqrt{3}$ is irrational using an argument like the one in Example 3. (*Hint:* All whole numbers can be expressed as $3k$, $3k + 1$, or $3k + 2$ for some whole number k.)

5. MATHEMATICAL REASONING Prove that $\sqrt{3}$ is irrational using an argument like the one in Example 4.

TEXTBOOK REFERENCES

Lesson 10.3

KEY WORDS

• binomial

CA STANDARDS

10.0

Linda has a square garden that is 10 feet by 10 feet. She wants to lengthen each side of the garden by the same amount x to create more space for her plants.

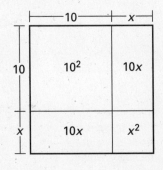

The area of Linda's new garden is $(10 + x)^2$. One way to find this product is to divide the garden into four pieces, as shown at the right. The area of the garden is the sum of the areas of the pieces.

$$(10 + x)^2 = 10^2 + 10x + 10x + x^2 \qquad \text{Add areas of pieces.}$$
$$= 10^2 + 2(10x) + x^2 \qquad \text{Combine 10x terms.}$$
$$= 100 + 20x + x^2 \qquad \text{Simplify.}$$

▶ **VOCABULARY TIP**
Recall that a binomial, such as $10 + x$ or $a + b$, is a polynomial with two terms.

You can use a similar method to find a formula for expanding the more general product of binomials $(a + b)^2$.

Example 1 *Finding a Formula for $(a + b)^2$*

Use a geometric model to find a formula for the product $(a + b)^2$.

▶ **Solution**

Start with a square having sides of length $a + b$. Divide the square into four pieces: an a-by-a square, two a-by-b rectangles, and a b-by-b square.

| Area of original square | = | Sum of areas of 4 pieces | Write verbal model. |

$$(a + b)^2 = a^2 + ab + ab + b^2 \qquad \text{Substitute algebraic expressions.}$$
$$= a^2 + 2ab + b^2 \qquad \text{Combine ab terms.}$$

▶ **Answer** The desired formula is $(a + b)^2 = a^2 + 2ab + b^2$.

Find the product.

1. $(x + 1)^2$ **2.** $(m + 5)^2$ **3.** $(2y + 3)^2$

You can also use geometric models to derive formulas for the products $(a - b)^2$ and $(a + b)(a - b)$.

Example 2 *Finding a Formula for (a − b)²*

Use a geometric model to find a formula for the product $(a - b)^2$.

▶ **Solution**

Start with a square having sides of length $a - b$. Add a border of width b along two sides of the square as shown.

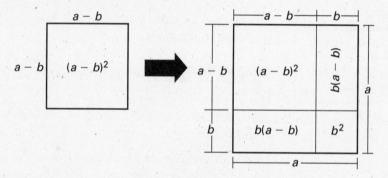

Area of original square	=	Area of original square and border	−	Area of border

$$(a - b)^2 = a^2 - [b(a - b) + b(a - b) + b^2]$$
$$= a^2 - [2b(a - b) + b^2]$$
$$= a^2 - 2b(a - b) - b^2$$
$$= a^2 - 2ab + 2b^2 - b^2$$
$$= a^2 - 2ab + b^2$$

▶ **Answer** The desired formula is $(a - b)^2 = a^2 - 2ab + b^2$.

In Exercises 4–6, find the product.

4. $(x - 2)^2$ **5.** $(k - 4)^2$ **6.** $(3u - 1)^2$

7. Use the formula for $(a + b)^2$ to derive the formula for $(a - b)^2$.
    (*Hint:* Replace b with $-b$ in the formula for $(a + b)^2$.)

Example 3 *Finding a Formula for (a + b)(a − b)*

Use a geometric model to find a formula for the product $(a + b)(a - b)$.

▶ **Solution**

Start with a rectangle having dimensions $a + b$ by $a - b$. Slice off a rectangle with dimensions b by $a - b$, and position it as shown to form an a-by-a square with a b-by-b square removed from one corner.

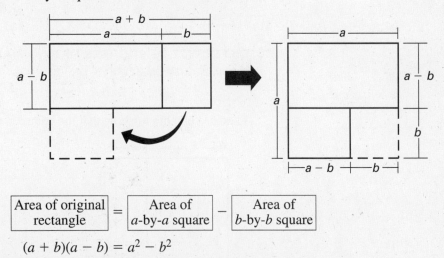

Area of original rectangle	=	Area of a-by-a square	−	Area of b-by-b square

$$(a + b)(a - b) = a^2 - b^2$$

▶**Answer** The desired formula is $(a + b)(a - b) = a^2 - b^2$.

Checkpoint ✓ *Products of the Form (a + b)(a − b)*

In Exercises 8–10, find the product.

8. $(x + 1)(x - 1)$ **9.** $(n + 7)(n - 7)$ **10.** $(5v + 2)(5v - 2)$

11. Find the product $(a + b)(a - b)$ by adding the areas of the two rectangles that form the original rectangle in Example 3.

The results of Examples 1–3 are summarized below.

FORMULAS FOR SPECIAL PRODUCTS OF BINOMIALS

Square of Binomial Formulas

$(a + b)^2 = a^2 + 2ab + b^2$ **Example:** $(x + 3)^2 = x^2 + 6x + 9$

$(a - b)^2 = a^2 - 2ab + b^2$ **Example:** $(2y - 5)^2 = 4y^2 - 20y + 25$

Sum and Difference Formula

$(a + b)(a - b) = a^2 - b^2$ **Example:** $(3n + 4)(3n - 4) = 9n^2 - 16$

Exercises

In Exercises 1–12, find the product.

1. $(x + 2)^2$

2. $(x - 1)^2$

3. $(x + 3)(x - 3)$

4. $(y + 10)(y - 10)$

5. $(c - 6)^2$

6. $(p + 15)^2$

7. $(2x + 1)(2x - 1)$

8. $(4m + 7)^2$

9. $(9n - 5)^2$

10. $(11 - t)^2$

11. $(1 - 8q)(1 + 8q)$

12. $(13 + 6r)^2$

13. DIFFERENCE OF SQUARES OF BINOMIALS Use the figure at the left below to show that:

$$(a + b)^2 - (a - b)^2 = 4ab$$

Then use algebra to show that the equation is true.

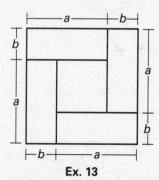

Ex. 13

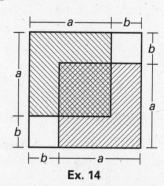

Ex. 14

14. SUM OF SQUARES OF BINOMIALS Use the figure at the right above to show that:

$$(a + b)^2 + (a - b)^2 = 2(a^2 + b^2)$$

Then use algebra to show that the equation is true.

15. SQUARE OF TRINOMIAL Use the figure at the left below to find a formula for expanding the product $(a + b + c)^2$.

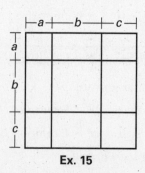

Ex. 15

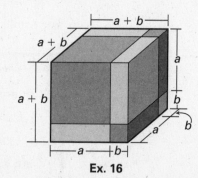

Ex. 16

16. CUBE OF BINOMIAL Use the figure at the right above to find a formula for expanding the product $(a + b)^3$.

TEXTBOOK REFERENCES

Lesson 11.3

KEY WORDS

• rational number
• terminating decimal
• repeating decimal

CA STANDARDS

1.0, 2.0, 25.0

WRITING RATIONAL NUMBERS AS DECIMALS

Recall that a rational number is a number that can be written as a quotient of two integers, such as $\frac{2}{3}$ and $\frac{-1}{5}$. By carrying out the division, you can write a rational number as a decimal.

Example 1 *Writing a Rational Number as a Decimal*

Write each rational number as a decimal.

a. $\dfrac{1}{20}$ **b.** $\dfrac{19}{33}$ **c.** $\dfrac{11}{15}$

▶ **Solution**

Using long division, divide the numerator by the denominator. (The remainder at each step in the division process is shown in **boldface**.)

a.
$$
\begin{array}{r}
0.05 \\
20\,)\overline{1.00} \\
\underline{0} \\
\mathbf{1}\,00 \\
\underline{1\,00} \\
\mathbf{0}
\end{array}
$$

b.
$$
\begin{array}{r}
0.5757\ldots \\
33\,)\overline{19.0000\ldots} \\
\underline{16\,5} \\
\mathbf{2}\,50 \\
\underline{2\,31} \\
\mathbf{190} \\
\underline{165} \\
\mathbf{250} \\
\underline{231} \\
\mathbf{19}
\end{array}
$$

c.
$$
\begin{array}{r}
0.733\ldots \\
15\,)\overline{11.000\ldots} \\
\underline{10\,5} \\
\mathbf{50} \\
\underline{45} \\
\mathbf{50} \\
\underline{45} \\
\mathbf{5}
\end{array}
$$

In part (a) of Example 1, notice that the long division process eventually results in a remainder of 0 and the division ends. The quotient, 0.05, is called a *terminating* decimal because it has a final, or ending, digit.

In part (b) of Example 1, notice that the long division process never ends because the remainders repeatedly cycle between 25 and 19. The quotient, $0.5757\ldots$, is called a *repeating* decimal because it consists of a block of digits, 57, that repeats endlessly. Another way to write $0.5757\ldots$ is $0.\overline{57}$ where the bar is used to identify the digits that repeat.

In part (c) of Example 1, notice that the quotient is also a repeating decimal, but it has a digit (7) that doesn't repeat in front of a digit (3) that does repeat. Another way to write $0.733\ldots$ is $0.7\overline{3}$ where the bar is placed only over the repeating digit.

A question to consider is why the decimal form of $\frac{1}{20}$ is a terminating decimal while the decimal forms of $\frac{19}{33}$ and $\frac{11}{15}$ are repeating decimals. The answer has to do with the fact that we use a base-10 system of numeration.

- For $\frac{1}{20}$, the prime factorization of the denominator is $2 \cdot 2 \cdot 5$, and all the factors of the denominator are also factors of 10. It is therefore possible to rewrite $\frac{1}{20}$ as an equivalent fraction with a denominator that is a power of 10, making it easy to write in decimal form:

$$\frac{1}{20} = \frac{1 \cdot 5}{20 \cdot 5} = \frac{5}{100} = 0.05$$

- For $\frac{19}{33}$, the prime factorization of the denominator is $3 \cdot 11$, and none of the factors of the denominator is also a factor of 10. This forces the remainder at each stage in the division process to be one of the numbers from 1 to 32, and as soon as one of these remainders appears a second time in the division process (after, at most, 32 steps), all the remainders encountered up to that point will be repeated in the same order over and over. This causes the digits in the quotient to repeat in the same order over and over.

- For $\frac{11}{15}$, the prime factorization of the denominator is $3 \cdot 5$, and one of the factors of the denominator is also a factor of 10 while the other is not. By introducing a factor of 2 in the denominator (and in the numerator), you can see why the decimal form of $\frac{11}{15}$ has a nonrepeating digit followed by a repeating digit:

$$\frac{11}{15} = \frac{22}{30} \qquad \text{Multiply numerator and denominator by 2.}$$

$$= \frac{1}{10} \cdot \frac{22}{3} \qquad \text{Write a product with one denominator a power of 10.}$$

$$= \frac{1}{10}\left(7 + \frac{1}{3}\right) \qquad \text{Express } \frac{22}{3} \text{ in mixed form.}$$

$$= 0.1(7 + 0.\overline{3}) \qquad \text{Write fractions as decimals.}$$

$$= 0.7 + 0.0\overline{3} \qquad \text{Distribute.}$$

$$= 0.7\overline{3} \qquad \text{Add.}$$

The preceding discussion suggests the theorem stated at the top of the next page.

THEOREM: WRITING A RATIONAL NUMBER AS A DECIMAL

When written in decimal form, the rational number $\frac{a}{b}$ (where a and b are integers having no common factors and $b \neq 0$) is either a terminating decimal or a repeating decimal. In particular, it falls into one of the following cases:

Case 1: If the prime factorization of b contains only 2's and 5's, then the decimal form of $\frac{a}{b}$ is a terminating decimal.

Case 2: If the prime factorization of b contains neither 2's nor 5's, then the decimal form of $\frac{a}{b}$ consists of a block of digits that repeats endlessly.

Case 3: If the prime factorization of b contains 2's and/or 5's as well as other primes, then the decimal form of $\frac{a}{b}$ consists of a block of digits that doesn't repeat followed by a block of digits that repeats endlessly.

Example 2 *Proving Part of the Theorem*

Prove Case 1 of the theorem stated above.

▶ **Solution**

Given $\frac{a}{b}$, write b as $2^m \cdot 5^n$ (where m and n are positive integers) and multiply numerator and denominator by $5^m \cdot 2^n$. Then $\frac{a}{b} = \frac{5^m \cdot 2^n \cdot a}{10^{m+n}}$. Since the denominator is now a (positive) power of 10, dividing by it will simply move the decimal point at the end of $5^m \cdot 2^n \cdot a$ to the left $m + n$ places. Since $5^m \cdot 2^n \cdot a$ has a finite number of digits, the decimal obtained by moving the decimal point $m + n$ places to the left will be a terminating decimal.

WRITING DECIMALS AS RATIONAL NUMBERS

If you are given a terminating or a repeating decimal, you can always recover the rational number having that decimal form.

Example 3 *Writing a Terminating Decimal as a Rational Number*

To write 0.52 as a rational number, use an appropriate power of 10 and simplify if possible.

$$0.52 = \frac{52}{100} = \frac{13}{25}$$

► STUDY TIP

In part (b) of Example 4, notice that the equation $999x = 344.79$ contains a decimal. Before solving the equation for x, you can multiply both sides by 100 to get an equivalent equation not having any decimals.

Example 4 Writing a Repeating Decimal as a Rational Number

Write each decimal as a rational number.

a. $0.\overline{47}$ **b.** $0.34\overline{513}$

► **Solution**

In each case, let x equal the given decimal. Multiply x by 10 raised to a power equal to the number of digits in the repeating block, then subtract x to eliminate the repeating digits.

a. Let $x = 0.4747\ldots$. Since there are two digits in the repeating block, multiply x by $10^2 = 100$, then subtract x.

$$\begin{array}{r} 100x = 47.4747\ldots \\ - \quad x = 0.4747\ldots \\ \hline 99x = 47 \\ x = \dfrac{47}{99} \end{array}$$

b. Let $x = 0.34513513\ldots$. Since there are three digits in the repeating block, multiply x by $10^3 = 1000$, then subtract x.

$$\begin{array}{r} 1000x = 345.13513513\ldots \\ - \quad x = 0.34513513\ldots \\ \hline 999x = 344.79 \\ 99{,}900x = 34{,}479 \\ x = \dfrac{34{,}479}{99{,}900} = \dfrac{1277}{3700} \end{array}$$

► STUDY TIP

In Example 5, notice that multiplying S by 2 shifts the terms of the sum just as multiplying the decimals in Example 4 by powers of 10 shifts the decimal points.

Example 5 Extending the Method of Example 4

CHESS A story about the invention of chess relates that a king in India was so delighted by the game that he offered to give the inventor anything he desired. The inventor asked that wheat be placed on the 64 squares of the board, with 1 grain of wheat on the first square and with each successive square getting twice the amount of wheat put on the previous square. Thus, there would be $2 \cdot 1 = 2$ grains of wheat on the second square, $2 \cdot 2 = 2^2$ grains of wheat on the third square, $2 \cdot 2^2 = 2^3$ grains of wheat on the fourth square, and so on up to 2^{63} grains of wheat on the 64th square. What amount of wheat did the king owe the inventor?

► **Solution**

The sum $1 + 2 + 2^2 + 2^3 + \cdots + 2^{63}$ represents the wheat owed to the inventor. To simplify this sum, let S represent it. Multiply S by 2 and then subtract S as follows:

$$\begin{array}{r} 2S = 2 + 2^2 + 2^3 + \cdots + 2^{63} + 2^{64} \\ - \quad S = 1 + 2 + 2^2 + 2^3 + \cdots + 2^{63} \\ \hline S = -1 + 2^{64} \\ = 2^{64} - 1 \end{array}$$

► **Answer** The inventor was owed $2^{64} - 1 \approx 20$ *quintillion* grains of wheat!

Without dividing, tell whether the decimal form of each rational number is a terminating decimal or a repeating decimal. Explain how you know. Then divide to obtain the decimal form the number.

1. $\dfrac{3}{8}$　　　**2.** $\dfrac{2}{3}$　　　**3.** $\dfrac{4}{15}$　　　**4.** $\dfrac{19}{25}$

5. $\dfrac{11}{18}$　　　**6.** $\dfrac{159}{200}$　　　**7.** $\dfrac{35}{44}$　　　**8.** $\dfrac{22}{39}$

9. Divide to obtain the decimal forms of $\dfrac{1}{7}, \dfrac{2}{7}, \dfrac{3}{7}$, and $\dfrac{4}{7}$. Look for a pattern in the decimals. Use the pattern to predict the decimal forms of $\dfrac{5}{7}$ and $\dfrac{6}{7}$. Check your answer by dividing.

10. Consider the rational number $\dfrac{a}{b}$ where a and b are nonzero integers having no common factors and where b has neither 2 nor 5 as a factor. Then the decimal form of $\dfrac{a}{b}$ has a block of digits that repeats endlessly.

a. What is the *minimum* number of digits that can appear in the repeating block? Find a rational number whose decimal form has this minimum number.

b. What is the *maximum* number of digits that can appear in the repeating block? (*Hint:* Your answer will involve b.) Find a rational number whose decimal form has this maximum number.

11. MATHEMATICAL REASONING Prove the second case of the theorem stated on page T19.

12. MATHEMATICAL REASONING Prove the third case of the theorem stated on page T19.

Write each decimal as a rational number.

13. 0.4　　　**14.** $0.\overline{6}$　　　**15.** 0.012　　　**16.** $0.708\overline{3}$

17. $0.\overline{153846}$　　　**18.** $0.3\overline{18}$　　　**19.** 0.555　　　**20.** $0.\overline{0099}$

21. In Example 5, suppose that the inventor of chess asked the king to put 1 grain of wheat on the first square and to triple the amount of wheat as he went from one square to the next. How much wheat would the king owe the inventor?

22. a. Simplify the sum $1 + 2 + 2^2 + 2^3 + \cdots + 2^{100}$.

b. Simplify the sum $1 + r + r^2 + r^3 + \cdots + r^n$ where $r \neq 1$.

TEXTBOOK REFERENCES

Lesson 12.6

KEY WORDS

• leg of a right triangle
• hypotenuse of a right triangle
• Pythagorean theorem
• tangent

CA STANDARDS

1.0, 2.0, 10.0, 25.0

▶**STUDY TIP**
Recall that the formula for the area A of a triangle with base b and height h is $A = \dfrac{1}{2}bh$.
For a right triangle, the length of either leg can serve as the base and the length of the other leg as the height.

ACTIVITY FINDING THE LENGTH OF A DIAGONAL OF A SQUARE

In this activity, you will find the length of a diagonal of the square shown. The square has sides of length a.

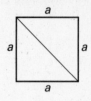

1 Begin by drawing a square having sides of length $2a$. How many squares of side length a fit inside the square you draw? Explain.

2 Mark the midpoint of each side of the square from Step 1, then connect successive midpoints to form an inner square surrounded by four congruent right triangles. Explain how you know that the inner figure is a square.

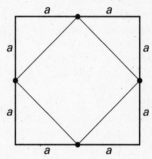

3 Let x represent the length of a side of the inner square. Write an equation relating x and a using this verbal model:

Area of outer square	=	Area of inner square	+	Combined areas of right triangles

3 Solve your equation from Step 3 for x. How is this number related to a square of side length a?

When a diagonal is drawn in a square, two right triangles are formed. If the length of a side of the square is a, then each triangle has legs of length a and a hypotenuse of length $a\sqrt{2}$. Notice that the sum of the squares of the lengths of the legs, $a^2 + a^2 = 2a^2$, is equal to the square of the length of the hypotenuse, $(a\sqrt{2})^2 = 2a^2$. This observation is a special case of the Pythagorean theorem.

▶**STUDY TIP**
The Pythagorean theorem applies only to right triangles, not to acute or obtuse triangles.

THE PYTHAGOREAN THEOREM

If a right triangle has legs of length a and b and a hypotenuse of length c, then the sum of the squares of the lengths of the legs is equal to the square of the length of the hypotenuse:

$$a^2 + b^2 = c^2$$

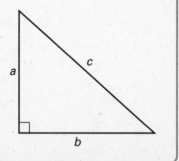

Example 1 *Proving the Pythagorean Theorem*

Prove the Pythagorean theorem by extending the argument developed in the activity to a right triangle with legs of length a and b and with hypotenuse of length c.

▶ **Solution**

Draw a square with sides of length $a + b$. Mark points that divide the sides into lengths a and b and connect them as shown. You obtain an inner square surrounded by four congruent right triangles. Let the length of the hypotenuse of each triangle (which is also the length of each side of the inner square) be c.

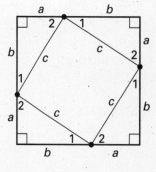

You can relate the areas of the squares and triangles as follows:

Area of outer square	=	Area of inner square	+	Combined areas of triangles

Write verbal model.

$$(a + b)^2 = c^2 + 4\left(\tfrac{1}{2}ab\right)$$

Substitute.

$$a^2 + 2ab + b^2 = c^2 + 2ab$$

Multiply.

$$a^2 + b^2 = c^2$$

Subtract *2ab* from both sides.

Therefore, for a right triangle with legs of length a and b and with hypotenuse of length c, $a^2 + b^2 = c^2$.

· ·

Checkpoint ✔ *Using the Pythagorean Theorem*

1. If you know the lengths of any two sides of a right triangle, explain how you can use the Pythagorean theorem to find the length of the third side.

2. A right triangle has legs of length a and b and hypotenuse of length c. Find the missing length.

    a. $a = 4, b = 7$ **b.** $a = 8, c = 10$ **c.** $b = 9, c = 11$

The Pythagorean theorem has many applications, especially in surveying, engineering, and astronomy where direct measurement of distances is difficult or impossible. One such application is finding the line-of-sight distance to the horizon for an observer positioned above Earth's surface. An approximate formula for this distance is derived in Example 2 on the next page.

READING TIP
The word *tangent*
comes from the Latin
word *tangere*, meaning
"to touch." A line that
is tangent to a circle
makes contact with the
circle at a single point.

Example 2 *Deriving the Line-of-Sight Distance Formula*

An observer is positioned x feet above Earth's surface. Assume Earth is a sphere with radius 4000 miles. If there are no obstructions blocking the observer's view to the horizon, about how far can the observer see? Express this distance d in terms of x.

▶ **Solution**

Draw a diagram showing a cross section of Earth with center C. The observer at O has a line of sight to the horizon at H, as shown.

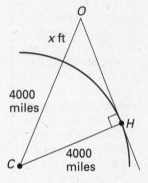

The line of sight, $\overline{OH}$, just touches, or is *tangent* to, Earth's surface. A theorem from geometry says that a line tangent to a circle is perpendicular to the radius drawn to the point of tangency. Therefore, $\angle H$ is a right angle and $\triangle OHC$ is a right triangle.

To find the distance $d = OH$, you can use the Pythagorean theorem. First you must express the lengths of both the leg $\overline{CH}$ and the hypotenuse $\overline{CO}$ using the same units of measurement. Since there are 5280 feet in 1 mile, Earth's radius is $4000 \cdot 5280 = 21,120,000$ feet. For convenience, let r represent this large number. Then $CH = r$ and $CO = r + x$.

$d^2 = (OH)^2$	Consider the square of the line-of-sight distance.
$\quad = (CO)^2 - (CH)^2$	Use the Pythagorean theorem.
$\quad = (r + x)^2 - r^2$	Substitute.
$\quad = r^2 + 2rx + x^2 - r^2$	Multiply.
$\quad = 2rx + x^2$	Simplify.

For an observer near Earth's surface, the value of $2rx$ is much greater than the value of x^2, so ignoring x^2 won't affect the value of d^2 much. So, $d^2 \approx 2rx = 2(21,120,000)x = 42,240,000x$, and $d \approx \sqrt{42,240,000x} \approx 6499\sqrt{x}$ feet.

To express d in miles, divide by 5280 and get $d \approx \dfrac{6499}{5280}\sqrt{x} \approx 1.2\sqrt{x}$ miles.

▶ **Answer** An observer positioned x feet above Earth's surface can see approximately $1.2\sqrt{x}$ miles.

· ·

Checkpoint ✓ *Using the Line-of-Sight Distance Formula*

3. LIFEGUARDING A lifeguard observes the water from an elevated chair that puts his eyes 10 feet above the beach. About how far can he see?

4. MATHEMATICAL REASONING Does doubling your height above Earth's surface double the distance you can see? Explain.

Exercises

In Exercises 1–6, a right triangle has legs of length a and b and hypotenuse of length c. Find the missing length.

1. $a = 6, b = 9$ **2.** $a = 5, b = 5$ **3.** $a = 12, c = 13$

4. $a = \sqrt{5}, c = 3$ **5.** $b = 21, c = 25$ **6.** $b = 2, c = 2\sqrt{10}$

7. MATHEMATICAL REASONING Many proofs of the Pythagorean theorem exist. The diagrams below show two ways to subdivide a square with sides of length $a + b$. Use the fact that the square's area is the same in both diagrams to write a proof of the theorem.

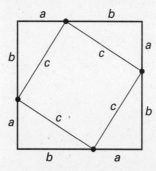 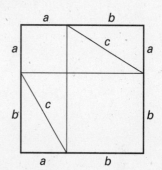

8. MATHEMATICAL REASONING The diagram at the right suggests yet another way to prove the Pythagorean theorem. Write a proof.

9. GEOMETRY Use the Pythagorean theorem to find the area of an equilateral triangle with sides of length a. (*Hint:* Draw a line through one vertex of the triangle and perpendicular to the opposite side.)

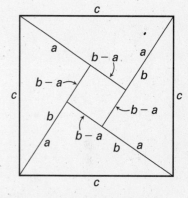

10. FERRIS WHEEL At the top of the Ferris wheel on Santa Monica Pier in Santa Monica, California, you are at a height of 86 feet. How far can see from that vantage point?

11. TALLEST BUILDING The observation floor of the tallest building in Chicago is 1353 feet above the ground. The distance between Chicago and the opposite shore of Lake Michigan is about 54 miles. Is it possible to see the opposite shore of Lake Michigan from the observation floor of Chicago's tallest building? Explain.

12. HOT-AIR BALLOONING While hot-air ballooning over the Great Plains, you notice that there are wheat fields in all directions as far as your eyes can see. If the balloon is floating 2000 feet above Earth's surface, about what area of land that you can see is covered in wheat? Explain.